The Untouchables

Second Edition

by
Brian Shul
and
Walter Watson Jr.

MACH 1, Inc.
Chico, California

MACH 1, Inc.
PO Box 7360, Chico, CA U.S.A. 95927

Copyright © 1993 Brian Shul and Walter Watson Jr.
Photographs by Brian Shul, Copyright © 1988 Brian Shul, except as follows:
p. 38, 181, 204 Copyright © 1993 Dino Corbin; p. 66, 91, 107 John Gaffney

Images provided by American Design And Marketing
Art Director: Lindy Hoppough
Set in 15/18 Adobe Garamond

No part of this publication may be reproduced or used in any form or by any means — graphic, electronic, or mechanical, including photocopying, recording, taping, or information storage and retrieval systems — without written permission of the publisher.

Library of Congress Catalog Card Number: 93-078250

ISBN: 0–929823–12-5

First printing, December 1993

Printed in Singapore by Craft Print Pte, Ltd.

Dedication

. . . For all those who, without compromise, kept her flying.

Foreword

In early 1992 Mach 1, Inc. introduced the book *Sled Driver: Flying the World's Fastest Jet*, by Brian Shul. Though we had high aspirations for this one-of-a-kind work, we couldn't predict the overwhelmingly positive response we received from readers all over the world. First with cards and letters, then by telephone, everyone, it seemed, wanted to share with us their pleasure with the book. We read and filed each card with great joy.

After receiving over fifty domestic and foreign book reviews, all praising Brian Shul's work, Mach 1, Inc. was notified that *Sled Driver* had been awarded two prestigious Golden Giorgi Awards. Just six months after the book's release, our readers' comments were validated as The Writers Foundation of America had voted our book best in its class in content and in publishing style.

As the letters of thanks and congratulations continued, we soon realized that many of you wanted more from Brian Shul about this magnificent airplane. Thus, the idea for the book you now hold was born.

Creating a new and original book, also centered on the SR-71, was no easy task. Rather than a sequel, or simply more of the same, Mach 1, Inc. and Brian were committed, as always, to producing a totally unique work; one which could stand on its own, yet make a good complementary volume to the popular *Sled Driver*.

Now, for the first time, *The Untouchables* echoes the voices of those proud people behind the scenes who were so integral in keeping the jet flying. Also, Brian's backseater, Walt Watson, brings to print the other half of the flying story in this two-man airplane. For the past year, they have worked diligently to bring their unique SR-71 experiences together in a format that they hoped would captivate the reader. After reading this book, I'm sure you'll find they have succeeded.

To all of you who took the time to write to us,
we read 'em all and loved 'em all – THANKS!

— Paul P. Farsai —
Publisher

Acknowledgments

My sincere appreciation to the following individuals who helped me in the construction of this book: Paul Mellinger and Bill Umble who were instrumental in providing names for interviews; Don Nolan and Marta Bonmeir of the NASA Flight Test Center; Michele Briggs from the Castle Air Museum; Base operations officer Bill Hayes at Beale AFB; Colonel Walt Watson at Maxwell AFB; and Paul Farsai of Mach 1, Inc, for faith in the concept, and having the patience to allow me to follow it to completion.

Contents

Foreword 11

Introduction 19

CHAPTER 1   **The Gathering Storm** 31
       The Critical Force 36

CHAPTER 2   **Iron Negotiations** 40
       The Critical Force 47
       Distant Thunder 52
       The Critical Force 55

CHAPTER 3   **The Eldorado Trail** 57
       The Critical Force 61
       By Dawn's Early Light 66
       The Critical Force 71
       Morning Passage 77
       The Critical Force 86
       Post Strike Plans 92
       The Critical Force 97

CHAPTER 4	**The Mission**	106
	The Critical Force	127
	Message in the Med	133
	The Critical Force	138
	Dinosaur on Patrol	148
	The Critical Force	154
	The Critical Force	160
	Across the 'Line of Death'	167
	The Critical Force	175
	This Boom's For You	180
	The Critical Force	184
	The Critical Force	186
CHAPTER 5	**The Final Run**	188
	The Critical Force	198
	Mission Accomplished	204
	The Critical Force	207
	Epilogue	208
	The Force	211
	Glossary of Terms	212

Kelly Johnson

" *I have known what I wanted to do ever since I was 12. There has never been any change since that time in my desire to design airplanes.*"

Kelly Johnson, 1910-1990

Introduction

The SR-71 was like no other aircraft ever built. Born from the genius of designer Kelly Johnson, the Blackbird served this nation for a quarter of a century with distinction. Initially designed as an interceptor in the early 1960s, the A-11 and YF-12 models validated Kelly Johnson's concept that he could build a plane that would sustain speeds of Mach 3 and better. The later strategic reconnaissance model, the SR-71, was the only version incorporated into the operational Air Force inventory.

Based at Beale AFB, California, the plane also eventually flew out of two overseas locations: RAF Mildenhall, England, and Kadena Air Base, Okinawa. It served six different presidents and saw action on hot and cold war fronts alike.

Because it was designed to fly higher and faster than any other air-breathing jet, the aircraft had to be constructed to withstand extreme temperatures while in flight. Also required were two engines capable of accelerating the jet to speeds of three times the speed of sound and keeping it there for long durations.

The achievements of Kelly Johnson and his team of engineers, in mating airframe with engines and inlet design are, today, legendary. The jet held every speed and altitude record when it entered Air Force service in 1966 and still held them when it was prematurely retired in 1990.

To operate and maintain such a remarkable and complex aircraft as the SR-71, required a host of remarkable people performing many complex tasks. Their dedicated efforts and own steadfast performance over the years rivaled that of the aircraft itself.

I was fortunate enough to know some of these people when, as an Air Force pilot, I was assigned to fly the SR-71. Crewed with Walt Watson, we flew together on many interesting reconnaissance missions over a four-year period. Our unique experiences with this incredible aircraft will stay with us for the rest of our lives. When the program was terminated in 1990, we were sad, but gratified, that we had been a part of something so few could claim. More people have stood on the top of Mt. Everest than have piloted the amazing SR-71.

I was inspired to write a book describing what it was like from my perspective in the front seat, and thus my first book, *Sled Driver*, was created. I knew there was so much more to the story than could be told in one book and was extremely happy when my publisher showed me the letters of appreciation from fascinated readers.

When Mach 1, Inc. approached me about a second book on the SR-71, I was enthused, but firm in my resolve to create something new. For me, the project had to have something special; something uniquely different from our first book.

The basic idea for a new book was to take the reader into the cockpit of the SR-71 on an actual mission, now that we could talk in more detail about such things. Additionally, my backseater, Walter Watson would add his input this time, giving the reader more than just the pilot's perspective. I suppose we could have sold the book on this premise alone, but I needed something more to inspire me to write with the same passion and energy that I felt with *Sled Driver*. I procrastinated, and few words were written.

Then a miracle happened. I don't know what else to call it — simply one of those memorable, unexpected events in life which occur at precisely the right time, in precisely the right manner.

A man named Paul Mellinger walked into my photography studio one day, with one of my books in hand, requesting my signature. I vaguely remembered the man as one of the Lockheed people who had worked with the SR-71 program at Beale. The most I could recall ever having dealt with him, was when Walt and I had gone to his office upon completion of our training to receive our lapel pins and desktop models of the SR-71. I had not seen the man in over four years and did not realize he had retired in the local area. I was, of course, quite pleased with his kind assessment of my book and remarked that I might even do another, but I wasn't sure. Then, unprompted, the man began to talk.

His words flowed like a fine wine, rich with recollections of the joy and little known nuances of the program which had occupied nearly a third of his life. I was fascinated with his broad knowledge of so many aspects of the SR-71 program. He spoke too, of his sadness at how the program was terminated, and how little people would ever know or understand about what it took to keep that program running so successfully for so long. He spoke of others who had worked with him, who had also retired in the local community, and how frustrated they felt, having to keep silent for all those years, now feeling somewhat forgotten. There was no bitterness in his voice, only the deep felt sense of loss for the commitment to excellence which that black jet had represented.

He spoke continually for two and a half hours, and as I listened, I could feel the intensity of his words. His story was not one of just airplanes, but of the human drama behind the jets. I found it fascinating to hear about the side of the program I had only seen glimpses of while flying the jet. Somewhere in that second hour, I realized I had been delivered a rare gift. The specialness I desired for my second book now sat before me.

When Mr. Mellinger left, little did he realize the seed he had planted. I was eager to see it grow and could never have imagined what a long and enlightening road I was about to tread in the creation of this book.

My journey began with Paul Mellinger himself. After several calls, I convinced him I was serious about interviewing him and he finally consented. I spent many hours at his home, pouring through boxes of SR-71 material he had saved, all meticulously sorted and labeled. As interesting as the memorabilia was, I found the man behind the bifocals even more fascinating. I had seen Paul a few times in the hangar when I flew the plane at Beale, always standing off to the side, always very quiet. I had no idea that he was the senior Lockheed field representative and was referred to as "Mr. Lockheed" due to his important position in the program. My impression had been that he was one of the engineer folks, performing some type of administrative duties for Lockheed, and he probably wasn't all that enamored with flying altogether.

Paul is an intensely private person, and only after several visits to his house, which was complete with HAM radio setup, did he speak more freely with me about himself.

Paul had joined the Navy while in college and, prior to the outbreak of World War II, served as a radio operator on a destroyer in the Pacific. He badly wanted to fly and was able to effect a swap of services to the Army Air Corps. During most of the war, he was an instructor pilot in the P-38 in Arizona, where he quickly became a squadron commander. With the invasion of Europe a certainty, all experienced pilots were needed, and on D-Day Paul found himself flying missions in the venerable P-47 from grass-covered fields in England. When Paul separated from the military in 1945 he, ironically, was discharged at Beale Field.

There was so much more than I had ever known to this quiet man, who had stood so many times in those hangars, always with a watchful eye on that black airplane. I soon realized that Paul Mellinger was the quintessence of the very special type of individuals who were at the heart of this program; the people who, over the years, kept this jet flying in a variety of ways. These were men mostly from that World War II generation who understood loyalty, discretion, the importance of a mission, and whose word was a bond of unwavering trust. I knew there were others like Paul who had a story to tell. Reluctant to talk about himself, Paul's greatest delight was in telling me of the many others who, he felt, had contributed so much to the SR-71 program. Soon I was off on an interesting trail of discovery.

The opportunity to talk with many "voices of the past" was an experience which not only enhanced my love for this phenomenal airplane, but gave me an even deeper appreciation for all those who had stood in support of me every time I climbed into the cockpit of this jet.

Before interviewing anyone else, I knew that any discussion of support had to start with the man who sat four feet behind me on every mission, Walt Watson. Walt was not only a consummate professional in the cockpit, but my good friend as well. I knew his perspective from the back seat would add greatly to the narrative of our missions, and I was pleased when he consented to take the time from his busy schedule to pen his thoughts.

The more people I interviewed, the broader became my fascination with these remarkable individuals. Walt and I had known many of them, of course, and even had dinner at some of their homes while TDY. It was illuminating now to hear them speak so many similar thoughts about their love for the airplane and the program as a whole. I had always thought that only the fliers could love the airplane in that special way. I was wrong.

During numerous interviews, the words *pride* and *dedication* continued to surface. These were more than just words to these people; they were concepts which were truly a part of their daily lives in the SR-71 community.

I listened quietly as Margaret, the Lockheed secretary, reminisced about how she felt watching the jet launch one day and had to stop talking to wipe tears from her eyes.

After enjoying dinner at his home, I watched as Doc, the engine specialist, proudly took a model off his shelf to show me — not of the plane, but of the J-58 engine, beautifully soldered from actual engine metals. When I asked him what the one weak part of the J-58 was, he hesitated thoughtfully, and then, somewhat apologetically, replied that he couldn't really think of one. He could not have given a better answer to one who had so often felt the reassuring thrust of that heavily muscled jet engine.

I marvelled at the group of women from the logistics center, who sat around a table and talked about the plane they had supplied parts to, with the kind of emotion and loving affection normally reserved for one's child.

I learned that Jim Cook, one of the tech reps, felt that no one really cared anymore about all that he and others had accomplished behind the scenes in the aerospace industry, but he is working on his autobiography, which will include his prominent work with the SR-71, so that his grandchildren will know something of his varied career as an aeronautical engineer.

There were magical moments too, like receiving a call from people I had never met — the Kelders — who were driving from Oregon to southern California, and wanted to know if they could stop by to get a signed copy of *Sled Driver*, because, after all, Sam Kelder had helped construct the fuselage on the first SR-71s, and, why yes, he'd be happy to talk about it. Sam came into my studio one day, sat down, and talked nonstop for three hours. He spoke like a man who had wanted to get it all out for years and had been waiting for someone to ask him.

There was the unexpected too, as Chuck Wiethoff's wife asked if she could sit in on the interview, anxious now to hear things Chuck had never talked about during his entire time with the program. She sat proudly for hours listening to her man speak humbly about his "minor" contributions to the program. She and I both knew better.

I watched Lew Williams, a longtime Lockheed employee become so animated with enthusiasm in talking about the jet that he had to stand up and walk around as he spoke. In stark contrast, was his discussion of the final days of the program, as he sat motionless and spoke in the hushed tones of one whose heart had been broken.

And in one hour I learned more from Bob Antilla about the space suit, than I had ever known in four years of wearing it.

I had the joy of listening to many others and realized, in a much deeper sense than I had ever felt before, that these people represented the true force behind all that we ever accomplished in the SR-71 program. So often, the public's view of the program was either one of unprecedented engineering feats, or the drama of men in space suits flying the world's fastest jet. Somewhere between the brilliance of Kelly Johnson and the glamour of a pilot's gloved hand on the stick stood the bridge of these men and women who made it all work: people who could take the concepts, and make them a reality. The little known, unheralded soldiers in the field who breathed life into a system that had no peer. Long silent, they needed to be a vital part of this book, their collective voice as important as the mission narrative itself. Since these people were, in a sense, with us on every mission, that is exactly where I put them — interspersed throughout the mission narratives.

Walt and I decided to write about our missions during the Libyan crisis in 1986, since they represented to us all that was meaningful about the airplane and the people in the Blackbird community.

The spring of 1986 was a turbulent time, as the threat of worldwide terrorist violence had reached a peak. Anti-American hostilities had gone on unanswered for years. Muammar Qaddafi had issued repeated threats to America specifically, and he indicated that those who crossed into the Gulf of Sidra would be crossing a "line of death." American servicemen were killed in a terrorist bombing in Germany, and the world seemed to stand hostage to the terrorists Qaddafi was sanctioning.

In a bold maneuver, President Reagan authorized the bombing of Libyan terrorist camps by F-111 fighter-bombers stationed in England. The two SR-71s also based in England would support the strike with aerial reconnaissance. With the exception of the steadfast support of Margaret Thatcher, the United States stood alone in this action, without the support of other nations, who feared recriminations. As part of the SR-71 detachment in England, Walter and I flew three consecutive days in support of this operation, something rarely, if ever, done in this jet. Military historians can long debate the effects of our military actions; suffice it to say that few threats were heard from Qaddafi for quite some time following America's succinct message.

In an attempt to keep close to the intent and feeling of those I interviewed for this book, I have intentionally left in certain idiomatic expressions, and rearranged words only for clarification. While there are so many people that go unmentioned in a book of this type, I feel strongly that those I did include are marvelous representatives of the many others who worked with this program. I have included their job titles and the number of years they were associated with the SR-71 program.

Though I never met Kelly Johnson personally, I felt compelled to include some of his words in a book such as this.

The cockpit narratives, taken from letters, diaries and phone conversations, were blended to place the reader directly into the action. I have also written the book so that, while the reading of *Sled Driver* may enhance the reader's understanding and enjoyment of the subject, it is not a prerequisite. There is also a glossary of terms to help the reader wade through common military jargon. And yes, there are still certain voids of classified information which will forever remain undiscussed.

The combining of two stories in this book — the mission, and the people behind the mission — will, I hope, lead the reader to the conclusion that they were one and the same.

Because of the many wonderful people I came to know in the construction of this book, I feel honored to have written it. You just couldn't touch these folks; and because of them, we weren't either.

— Brian Shul —

The Untouchables
1966-1990

Pilot/RSO

John Storrie/'Coz' Mallozzi
Gray Sowers/'Butch' Sheffield
Al Hichew/Tom Schmittou
'Pete' Collins/Connie Seagroves
John Kennon/Cecil Braden
Bill Campbell/Al Pennington
Pat Halloran/Mort Jarvis
Buddy Brown/Dave Jensen
Dale Shelton/Larry Boggess
Jerry O'Malley/Ed Payne
Don Walbrecht/Phil Loignon
Earle Boone/Dewain Vick
Tony Bevacqua/Jerry Crew
Jim Watkins/Dave Dempster
Ben Bowles/Jimmy Fagg
Larry DeVall/Clyde Shoemaker
Bob Spencer/'Keith' Branham
Brian McCallum/Bob Locke
Roy St Martin/Jim Carnochan
George Bull/Bill McNeer
Bob Powell/Bill Kendrick
Charlie Daubs/Bob Roetciseonder
Bobby Campbell/Jon Kraus
Abe Kardong/Jim Kogler
Nick Maier/Garry Coleman
Willie Lawson/Gil Martinez
Dave Frehauf/Al Payne
Jim Hudson/Norbert Budzinski

Pilot/RSO

Maury Rosenberg/Don Bullock
Joe Kinego/Roger Jacks
Al Cirino/Bruce Liebman
Jay Murphy/John Billingsley
Rich Graham/Don Emmons
Bob Crowder/John Morgan
Tom Alison/'JT' Vida
Buz Carpenter/John Murphy
Jack Veth/Bill Keller
Bill Groninger/Chuck Sober
'BC' Thomas/Jay Reid
Tom Keck/Tim Shaw
Dave Peters/Ed Bethart
Lee Shelton/Barry MacKean
Rich Young/Russ Szczepanik
Gil Bertelson/Frank Stampf
Rich Judson/Frank Kelly
Neven Cunningham/Geno Quist
Jerry Glasser/Mac Hornbaker
Maury Rosenberg/'ED' McKim
Rick McCrary/Dave Lawrence
Bernie Smith/Denny Whalen
Gil Luloff/Bob Coats
Les Dyer/Dan Greenwood
Bill Burk/Tom Henichek
Jim Jiggens/Joe McCue
Bob Behler/Ron Tabor
Stormy Boudreaux/Terry Newgreen

Pilot/RSO
Harlon Hain/'Butch' Sheffield
Tom Estes/Dewain Vick
Jim Shelton/Tom Schmittou
Dave Cobb/Myron Gantt
Tom Pugh/Ron Rice
Denny Bush/Phil Loignon
Randy Hertzog/John Carnochan
Dave Cobb/Reg Blackwell
Bob Cunningham/'GT' Morgan
Ty Judkins/Clyde Shoemaker
Jim Sullivan/Noel Widdifield
Bob Gunther/Tom Allocca
Carl Haller/John Fuller
Pat Bledsoe/Reg Blackwell
Mark Gersten/Lee Ransom
Harold Adams/Bill Machorck
Bob Helt/Larry Elliott
Al Joersz/John Fuller
Jim Wilson/Jim Douglass

Pilot/RSO
Jerry Glasser/Ted Ross
Joe Mathews/Curt Osterheld
Jack Madison/Bill Orcutt
Ed Yielding/Steve Lee
Brian Shul/Walter Watson
Duane Deal/Tom Veltri
Duane Noll/Charlie Morgan
Rod Dyckman/Tom Bergam
Mike Smith/Doug Soifer
Dan House/Blair Bozek
Tom Danielson/Stan Gudmundson
Terry Pappas/John Manzi
'Mac' McKendree/Randy Shelhorse
Larry Brown/Keith Carter
Steve Grzebiniak/Jim Greenwood
Tom McCleary/Blue Vardaman
Gregg Crittenden/Mike Finan
Don Watkins/Bob Fowlkes

The Untouchables

"... the enemy understood well the stalling tactics of 'continuing negotiations' which bought time for his forces and prolonged the conflict. When one side finds it necessary to discontinue fruitless talks, you, as fighter pilots, may become intimately familiar with the warrior's fundamental law of wartime negotiations, which is, you negotiate with the enemy with your knee in his chest and your knife at his throat."

MAJOR B. SHUL
CLASSROOM LECTURE
FIGHTER TRAINING COURSE
HOLLOMAN AFB, NM, 1982

CHAPTER 1

The Gathering Storm

FROM THE FRONT SEAT...

In the weeks preceding the raid on Libya, international tensions had been running high. Walt and I talked about some possible options our President might take concerning the increased level of sponsored terrorist attacks against the western world. We had been in England three weeks into our normal six week deployment and had flown four operational SR-71 sorties, all familiar routes covering areas to the north and east.

In the week prior to the actual strike, we began to notice an increased number of tanker aircraft arriving at Mildenhall, primarily KC-10s. I was beginning to get the impression that every KC-10 in the Air Force inventory just happened to drop in to Mildenhall for the week. This dramatic buildup in the number of planes on the ramp did not go unnoticed by the civilian populace either. The Brits are fanatical airplane watchers and are often seen lining the perimeter fence with binoculars in hands, making notes of various new tail numbers they observe. The observers around the base fence seemed to increase proportionally with the increase of jets on the field.

Normally, the military explanation for things of this nature was that the tankers were there in support of deployment exercises. True was the fact that the F-111s at nearby RAF Lakenheath had been doing an increased amount of flying in the previous weeks, but I knew that nobody gets that many tankers, from that many different units across the U.S., unless something more than just an exercise is in the making. Besides, the F-111s had been flying very little as more and more tanker aircraft lined the ramp at Mildenhall.

Walt and I had learned not to get too excited about anything though, as we had already gone through an increased alert condition two weeks prior. At that time the U.S. Fleet in the Mediterranean had taken a strong military posture in response to Muammar Qaddafi's repeated military threats. Too many times in the recent past, America had taken assorted abuse from hostile nations and done nothing, so there was no reason for me to believe this time would be any different.

FROM THE REAR SEAT...

I guess I wasn't as surprised as most people were when the actual bombing raid took place over Libya. Before volunteering for SR-71 duty, I had been one of the senior Weapons System Officers in the F-111 at Cannon AFB in New Mexico. While flying the SR-71 out of DET 4, I found many of my former F-111 students and friends now stationed just five miles down the road at RAF Lakenheath. Occasionally Brian and I would go over to the Lakenheath Club and talk some flying with the boys. Mildenhall was pretty much a MAC base, so you didn't get much of that there. Brian had known some of the '111 guys from his days of teaching at the Fighter Lead-In School, a necessary stop on every fledging fighter pilot's way to his primary fighter. When Brian and I went over to Lakenheath, we enjoyed seeing how experienced some of our former students had become. We both felt good to see some of the product that we had helped to form, in some small way, in the tactical flying force.

In the week preceding the Libyan raid, I noticed the '111 guys wanted to talk tactics with me, more than usual. Although they never gave me any details, I had a pretty good idea of what was brewing. There had already been some clashes between the U.S. Navy and Libyan MiGs, and the news was filled with increasing evidence that Libya was sponsoring terrorist attacks worldwide. I felt that the terrorist bomb which had killed U.S. servicemen at a nightclub in Germany was the last straw for our President.

When I had been flying with some of these guys back at Cannon, our unit had practiced many low level, night attack missions over the mountains and high desert of New Mexico. We had even deployed, one time, over to Egypt for an exercise to give us practical experience in a part of the world we would normally never see — unless we really were going to war. The experience proved to be invaluable in the development of good tactics for a desert environment.

The F-111 is a great platform for high speed, low level delivery of weapons. It has great range, carries a meaningful load of destruction, and can deliver accurate weapons on a target at 600 knots and higher. I knew the '111 was the one jet in the Air Force inventory that had enough range, and the nighttime low altitude capability, for the type of strike I surely felt was pending.

F-111s in arming area.

In the low-level environment, we understood that speed was life. The high speed makes it more difficult for the crew to hit the target but is necessary for survival. As I talked with these young crews, I saw in their faces such youth and enthusiasm. At some moment in our conversations, I realized I was the only one in the group with actual combat experience; the only one there who knew what it was like to be shot at. I guess I wanted them to be a little less enthused about the prospect of what they soon might be doing. But it really is the only way I would expect them to be.

We all had a laugh about the SAC proposal that B-52s be used to bomb selected targets if it came to that. SAC held pretty strongly to the idea that if you had electronic countermeasures on your aircraft, you were invisible. The wreckage of over a dozen BUFFs scattered across the Hanoi countryside had pretty much dissolved that theory for most of us. That magic black box stuff helped, no doubt about it, but we didn't want to bet our lives on it. One pilot in the group joked that the only way the B-52 could be effective in a heavily defended desert environment was if, after massive hits by SAMs and anti-aircraft fire, a big enough piece of aircraft wreckage happened to fall on the target. We all knew the BUFF was a good ol' plane in its environment, but each squadron had its own preferred tractics, and for the '111 guys, it was on the deck, as fast as you could stand it.

During the week, I had dinner at the homes of a couple friends at Lakenheath. Although no one told me anything about any upcoming operation, I found it easy to read their faces. At one home, during the saying of grace before dinner, I noticed that certain look on the face of my friend's wife that let me know they we were going to put it all on the line. There is that certain, detached look wives get that seems to say everything. It is a look that we all hated to see, but one that represented everything we would be fighting for, if called to battle. I played with their children after dinner, and in a way, it helped calm my own thoughts about what role I might play in the scenario I now felt certain was coming.

Most of the guys in the squadron at Lakenheath were fairly young. My experience told me that launching an F-111 attack into a country such as Libya would likely result in a predicted loss of two to three jets.

I remembered having to deal with the realities of these type of numbers with my own wife during many years of F-4 and F-111 flying. During my '111 days, I had told her that if my aircraft had taken a hit and I was missing, not to expect my return. At 620 knots, you are travelling at 1047.8 feet per second, and it doesn't take much to rip a large hole in the aircraft; and at that speed there are few pieces left, of anything. I had told her not to ask to see the body. As I sat there with these young families, I wasn't sure these guys had discussed such things with their wives, and I thought of my family back in California.

As the days went by, Qaddafi didn't seem to grasp the big picture and issued even more inflammatory remarks against the United States through the media. I knew it was just a matter of time now. The DET commander told Brian and me, and the other SR-71 crew, not to go anywhere, to be on a standby status. This was quite irregular.

I tried to keep myself occupied. I went with Brian to shoot some baskets at the gym, wrote letters, and even worked on my Air War College paper. Time seemed to move in slow motion. I couldn't help thinking about my buddies five miles away and the type of mission planning they must be doing. I had a lot of confidence in these guys though and knew some of them to be the top WSOs in the Air Force.

I also knew one of the pilots very well. He was now one of the experienced 'old heads' in his squadron, but I remembered him as a "new weenie," an inexperienced lieutenant at Cannon, who had been crewed with me so I would keep him out of trouble. Explaining the importance of the guy in the right seat to the pilot was always more difficult than simply showing him. I was afforded this opportunity one time when he and I were flying in the '111 at Redflag. We were coming out of the target area as number two in a two-ship formation, flying 6,000 feet apart in tactical spread for mutual support. Being inexperienced, he had gotten us well out of position, and sure enough, an "enemy" F-15 appeared out of nowhere, and tries to jump our leader, who is now assumed to be alone since we had gotten so far wide, and aft of our normal position. We still carried the gun in the '111 in those days, so while the "new guy" to my left is still trying to find his leader, I pick up the F-15 electronically, vector us over to his six o'clock position, lock him up, and with the twin

Lakenheath F-111 in shelter.

tails of the Eagle jet in our gunsight, calmly point out to my pilot that bagging an F-15 before he can "kill" our leader would be considered very good etiquette at this point in the mission. We were always very good friends after that, and with enough time in the '111 he had become an excellent pilot who was now teaching the others.

I thought, too, of my good buddy who shared an office with me at Cannon, when we were responsible for giving the evaluation checks to the WSOs. He and I shared a lot of ideas about tactics then, and I was now glad that someone with his expertise was in on the mission planning at Lakenheath.

Just waiting, that was the worst part.

In 1969 a British Royal Navy F-4 Phantom II traversed the distance between New York and London in the, then record time of 4 hours and 46 minutes. In 1974 the SR-71 set a new record for the same distance of 1 hour, 55 minutes and 42 seconds. Forty-seven years earlier, Charles Lindbergh had flown approximately the same distance in 33 hours and 29 minutes.

THE CRITICAL FORCE . . .

People always thought I was the "secretary" to the Lockheed reps, but in fact, my official job title was Administrative Engineering Assistant, and typing was the least of my jobs.

I came to Beale in 1965 and was working in another job when I saw the first few Blackbirds arrive. A new squadron was starting up and there was a definite air of mystery surrounding it. I was never an airplane person at all, but when I first saw that plane, I fell in love with it. It had such a personality all its own. You couldn't look at her and not feel something that moved you inside. I don't know how the crews looked at her, but to me it was something very alive.

Working with the Lockheed reps was exciting right from the beginning, and it pretty much stayed that way the whole time. We realized we were working with the premier aircraft in the world, and that never changed the entire time I was there. The dedication of those people in the early days was, well, something I'll just have to tell you about because you don't see that sort of thing anymore.

I left my old civil service job to become a Lockheed employee, and stayed in that job from the beginning of the first SR-71 squadron until the very end of the program, over twenty years later. I was in weekly contact with the folks down at Burbank as I would be sending out reports constantly. Typing reports in copies of eight on that old clunker of a typewriter gave me sore fingers at the end of the day, but I was just so proud to be a part of all that was going on with this plane. I don't remember ever being bored in my job. We frequently worked on weekends in those early days and I don't remember anyone ever complaining about that.

Initially, I typed up all the flight reports. Paul Mellinger, a great man to work for, was really good about keeping the folks at Burbank apprised of all the activities with the plane at Beale. I also handled all the testing charts. There was a lot of testing going on in those days and many, many charts needed for briefings and analysis. I also started keeping track of aircrews' time in the plane for making out certificates of hours accomplished. After being there awhile, I took it upon myself to devise a system to help locate classified materials kept in the safe. Eventually, my job became so multi-faceted that no one was quite sure exactly what I did.

Oh, I wish we could have had computers back then; it would have been so nice. But we didn't complain; there was such a pride in everyone's effort. You know that was back in the days when you didn't have to plaster the word *pride* across the side of buildings on base to remind everyone what it was they were supposed to have. No one needed to explain to us the importance of this airplane; we all just felt it.

The really unique thing was that I got to work with a basic group of people that didn't change much in twenty years. Now, I guess that could be bad in some circles, but in this case it was terrific. I never saw so many people from such diverse backgrounds work together so well and have such respect for each others' ideas. I was the only woman in that office the whole time, and these men were such gentlemen, very respectful of me as a person, and they were that way with each other. There were many disagreements, to be sure, and for the first five years I didn't know if Paul Mellinger had a sense of humor, but always there was this underlying belief in the importance of the program which held us together. Everyone felt that their part was very important.

My job changed somewhat as soon as some of the field reps started going TDY, initially to set up the DET in Okinawa. This meant, slowly but surely, I was picking up some of their administrative duties. Then of course, when they returned, they saw how efficiently it was being done so were happy to "allow" me to continue with those additional duties. I never complained, though; they had so many other important issues to work. I honestly would have worked with those people for half the pay. I never did get to go TDY, though; I always asked Paul and he'd always tell me "next time."

I did get to tour the plant a couple times though, and that was really something. I also had my one big moment in the simulator. I was terrified that I might hurt something. To me it was like the real plane, so I couldn't bring myself to realize it was just a simulator and that anything I did wasn't going to hurt the actual plane. I know it's silly, but I just sat there and stared at all those gauges. I think the guys were a little disappointed that I didn't fly the thing. It was an honor just to sit there in that seat where so few got to sit. I don't think the crews felt quite as enamored with the sim as I did.

Over the years I would get to see all the different crews come by since they had to pick up their certificates and SR-71 pins from us. These guys were great individuals. That never changed over the years.

I loved them for taking me out to see the plane on occasion, and when I was invited to observe a launch, I just loved being out on the flight line. It was very exciting watching all the people in action. When the nose of that jet came rolling out of the hangar and the crew chief saluted, it gave me goosebumps. I always asked them to roll down the window in the mobile car so I could hear the sound, that wonderful sound. As it taxied out, it truly did seem like a living, breathing entity. I always thought that little red flashing light on the underside of the fuselage was like the plane's heartbeat.

I guess we all felt a little possessive about the plane. If I knew one was down on an emergency landing somewhere in the country, I would feel like one of my children was lost and needed to get back home. That was our baby, and it was the world's best, and we wanted to see it back at Beale performing its mission; a mission we all felt was of the highest priority.

You know, many of us from those days still get together for lunch, and I guess as time goes by we've begun to realize just how special and talented the people on this program were.

Looking back, I feel that the plane served with the highest distinction. Just consider all the different political situations in the world from 1966 until 1990, and realize that during many of them this plane played a key part in gathering information for our leaders. That's why it was so hard at the end, seeing the program terminated when it was so vital and so healthy.

You know, maybe you should never ride a program right to the end, because you have to see the destruction of all that you created and watch a new regime take over that has little respect or remembrance of all that went before it. It was the saddest day of my life when they terminated the program. We had been like a family there. One by one, my friends and fellow workers left or retired. There was no ceremony for them, no big farewell speeches. One day Paul and Lew simply weren't there anymore, and it was sad, and it was never as good. I stayed to the bitter end. I was the last SR-71 Lockheed employee there. I turned out the lights and closed the door and have never been back.

That period was the fastest and most satisfying twenty years of my life. Sometime later, I saw the plane, retired, on display. To me, it was all wrong seeing it like a stuffed animal, quiet and motionless, with no feeling. It wasn't the jet I had known . . . there was no flashing heartbeat . . .

— *Margaret Martin* —
Administrative Engineering Assistant
23 years

" *I've been around aviation for over thirty years and this was by far the most awesome and most beautiful airplane ever built. They buried it while it was still alive.*"

CIVILIAN EMPLOYEE, DET 6

CHAPTER 2

Iron Negotiations

From the Front Seat . . .

14 April, 1986
Day One
0930

Finally, the secure phone in my room rings, and I know it is the commander or operations officers at the DET. I am told to get Walter and come into the squadron building, and we are not to wear any uniforms. I feel like we finally may be getting some mission tasking and am anxious to get briefed.

Walt and I walk separately to the DET, only about a half mile away. We took a circuitous route so as not to arouse suspicion, but I doubted seriously at the time if anyone really noticed us, since large numbers of British "tail counters" were already perched around the perimeter fence. They were indulging themselves with their usual detailed observations of every insignia on the newly arrived tankers now parked on every available piece of concrete around the field. We arrived for the briefing in a variety of sweat pants, jeans, and cap and gloves, since April in England has nothing to do with warm weather.

Along with the other SR-71 crew, we sit around a large table, the door is closed, and we are told we are about to get a Top Secret briefing. The room is very quiet as we are shown maps drawn with routes to Libya, and we are informed that the 48th Tactical Fighter Wing at Lakenheath will be striking selected Libyan targets within hours. No more posturing, no more exercises; this is a real mission with real bombs.

No matter how intense our dislike had been for all that Qaddafi stood for, it still sent a little shockwave through my mind to realize that American jets were really going to respond in anger. Part of me said, "It's about time," and another part of me feared what this type of operation could escalate into. Our greatest satisfaction in our type of work was that through a strong military posture, actual armed conflict could be avoided. It was with a sad resignation that I realized we had come to the bomb dropping option, and yet, sixteen years as a fighter pilot had shown me that, regrettably, there were still people like Qaddafi who understood our resolve for peace only when iron bombs were falling in his backyard. I knew he would not expect this response from a nation which had long since lost its "big stick" reputation. For years, the U.S. had condemned verbally but continued to watch helplessly as terrorist bombings around the world had claimed the lives of many civilian and military people alike. From past dealings with the West, Qaddafi had every reason to believe he could continue to get away with sanctioning, harboring and training the very terrorists who were growing in numbers, and he knew our response would likely be a quiet one, militarily. He was about to get a very rude wake-up call.

We looked over the route we would fly with great interest. At first I couldn't believe France would not issue permission for overflight by our aircraft. Then I thought about it for a minute, and it really didn't surprise me. This meant quite a long flight for us and excruciatingly long for the fighters.

The first role the SR-71 would play in the operation would be post-strike reconnaissance. Our DET would launch both SR-71s, one as the primary jet and the other as backup. Walt and I began our mission planning and got down to the business of dealing with the particulars of the mission, now officially designated Operation Eldorado Canyon. My first concern was locating the exact refueling points and the divert bases. The course route itself was fairly simple, but all the contingencies needed to be addressed.

I get briefed on how some of those KC-10s on the ramp will be refueling us, in formation with our normal KC-135s. That makes me happy since the '10 is a very stable platform for us to refuel from and it carries plenty of gas. Then they tell me the refueling altitude, and I'm not so happy. Due to civilian carrier airspace restrictions in the Mediterranean, the best we can have is a block altitude of 27,500 to 33,500 feet. Those altitudes would make it tough on the jet to stay on the boom, heavyweight. Normal refueling altitudes of 25,000 feet required the lighting of one afterburner as the jet became heavy with fuel near the end of the track. These higher altitudes would likely require both 'burners cooking, and I tried not to think too much about what that was going to feel like. The 500-foot suffix to the base altitudes also told me that we were in uncontrolled airspace, and it was our responsibility to "see and avoid" other planes now. This was a far cry from the specifically designated refueling tracks and altitudes we were used to in the States, where we were afforded traffic separation via radar controllers. Such is the nature of combat operations. We simply do what is necessary to get the job done. I could only hope for excellent weather in the Mediterranean.

Walt and I worked separately during the initial mission planning, as he tackled the sensor activation points and I concentrated on the flying portion. Then we talked together and went over numerous questions: How much fuel do we need to get home subsonically if we have a problem? What are the restrictions about landing in Spain or France if we really have to? (I had already made up my mind that I would do everything possible to land anywhere but in France.) Where can we safely land once into the Mediterranean? How long is the runway in Sardinia? Is Malta an option? What is our programmed fuel supposed to be coming out of the target area? What's plan B if our primary tanker can't pass gas? What's plan C? Where will the U.S. Navy fleet be located, and what type of assistance can we expect if we have to bail out feet wet? And finally, what threats can be expected in the target area? Even though Walt and I were scheduled to be the backup jet on the first day, we planned as if we were going all the way to the target area so we'd be better prepared for that contingency.

From the Rear Seat . . .

Day One
0930

The obvious increase in KC-10 aircraft on the ramp was the final clue for me. I deduced that they would be used to refuel the '111 strike package and maybe even to refuel us too. As I finished making a small breakfast, we got the call. There had been no contact from my friends at Lakenheath for several days, and I knew they must be readying themselves. From the building where we were housed, it was a short walk to the DET, and I felt a slight apprehension about all we were going to hear when we got there.

Finally, we were told that F-111s from Lakenheath, with some support from EF-111s at Upper Heyford, would launch a strike on Qaddafi's bases in Libya. The room became very quiet as we

The route. Hash marks are refueling tracks.

were briefed on the particulars. At first light, the SR-71 would make a pass over the target areas to get a bomb damage assessment. Both SR-71s at the DET would be launched, one as a backup in case the primary aircraft had a problem prior to the target area. Our routing was primarily dictated by the fact that France said no to our request for overflight. That meant the whole strike force, including us, would have to make the long trip around Portugal, through the Strait of Gibraltar, and then into the Mediterranean. Flying over French airspace would have greatly shortened the route for everyone, but the politics of the situation was not our concern, and we had much mission planning to do.

Our intelligence officers gave us some particulars concerning the threats we could encounter and a general overview of the route and tanker information. While the planners were busy drawing up the actual maps, Brian and I began discussing the mission in general, as did the other crew. The route, though fairly long, was pretty straightforward, and most of the critical turns and speeds were all in the target area.

The SR-71 was unique in that as much of a cohesive unit as the two people inside it needed to be, the duties in each cockpit were radically different. For the pilot, each mission was essentially the same concerning his primary duties. Though the specific route or refueling points would change, his job remained unchanged: fly the airplane, get the gas, take off and land safely. For the guy in the back, each mission could mean a different array of tasking depending on the routing, the type of sensors, the threat, and specific communications requirements. The cockpit itself could even be different depending on certain types of equipment.

Regardless of the route or type of mission we were flying, I always considered my primary job in the jet to back up the pilot in any way he needed. I figured if he needed it, then it was in my best interest to drop everything else and help him. That jet was my ticket home, and if it was sick, getting the pictures really didn't matter at that point. Since some gauges in my cockpit were the same as the pilot's, I could offer some assistance to him with certain parameters such as speed, headings or aircraft attitude. He, on the other hand, had nothing in the front seat that enabled him to assist me with such critical areas as sensors, the navigation system, defensive systems or camera settings. So for the guy in the back, each mission was a mental test, blending in the priorities of helping the pilot with the jet and accomplishing the varied actions required for mission success in the back.

By the time we began planning for this mission, Brian and I were a finely tuned crew, in that we both had a good understanding of each other's duties, and both had learned some compromises in order to help the other guy. This was a critical factor, and given the level of experience and strong egos within the SR-71 crew force, this smooth blend of mutual respect and compromise didn't always manifest itself. At the speeds we were flying, and the geography we were penetrating, there was no time for a discussion in flight, when for example, I might detect a navigational error and tell Brian to turn immediately. He knew to turn. And when he told me to start reading him an emergency page in the checklist, I knew he saw something critically wrong with the plane, and it was time to stop everything else and work that problem. We had that kind of trust in each other. It did not come easy, we earned it through many hours in the simulator and sitting on the floor of my den, late at night going over the checklist again and again until we really understood how the other guy was thinking. By the time we found ourselves planning for a flight toward Libya, we were both on the same page and it made all the difference in being able to deal with all the unexpected things which invariably occurred in flight. The trust we had in each other's abilities enabled us to divide up the planning into areas more pertinent to each cockpit, and saved us time. When we were done, we went over everything and felt good about our game plan.

The senior crew which was planning for the primary route, I believe, did not have the mutual respect or trust which Brian and I enjoyed as a crew. They had many disagreements and their mission planning took an excessive amount of time. I felt there already was enough to worry about.

One big difference in the mission would be the joint use of KC-135Q and KC-10 tanker aircraft. We normally refueled with the '135Qs. They carried the special communication and ranging equipment which helped us locate them in the big expanse of sky. On this mission, the "Q"s would

A finely tuned crew.

be in formation with the KC-10s. The "Q"s would still give us the necessary ranging and azimuth information, but the larger KC-10s would give us the gas, thus everyone would be able to make the long trek home with ample fuel. To further complicate the refueling, we would be hooking up at higher than normal altitudes. Brian and I both knew this would be tricky, but if the weather was good, it wouldn't be too bad. The jet really didn't like being that slow, that heavy, that high, while topping off with fuel, and we were definitely going to need every drop of gas.

Brian and I also talked a great deal about divert bases along the route. This was a critical subject, as we needed to act quickly and correctly in the event of an emergency, so as to put that plane down in safe territory. Basically, once we got into the Med, everything to our left was good, and everything to our right was bad. The team of planners drawing up the route maps circled bases along the route considered appropriate for emergency landings. Due to the political sensitivity of this operation, there were suddenly very few airfields we could count on. We had a good record of never having to land away during an operational sortie and we wanted to keep it that way.

We also talked about the threat. We knew that, following a bombing, those folks were going to be pretty unpleasant. Brian and I both knew that Qaddafi would love to get a piece of this jet in his hands just to show the world or, better yet, parade the crew members in front of the TV cameras. I joked with Brian that this was the one place we had ever flown over that I would better be able to blend into the populace than he would. Most of the places we flew over, if shot down, I would have stood out like a lump of coal in an egg basket. It was better to joke like that than to actually imagine standing in your space suit with your parachute in your hand, on Libyan soil.

We had been thoroughly briefed on all the Soviet equipment that the Libyans had, and it was certainly enough to get our attention. Unlike many other sorties we had flown, where we could be reasonably sure of not being fired upon, we could not feel that way now. After having been bombed, the Libyans would have nothing to lose by firing a barrage of missiles, even if they missed.

Brian and I discussed the DEF, though we both knew our best defense was altitude, speed and a turn if needed. The Libyans also had the MiG 25, so the air threat had to be considered. After the way a couple of Libyan pilots had made threatening passes against a couple of our F-14s and then cleverly trapped themselves in the Tomcats' gunsights, I wasn't too impressed with Qaddafi's pilots, but only a fool underestimates the enemy, going into battle, and I reviewed all the DEF settings for airborne threats.

I always felt that if something was going to get us, it would be a surface launched missile. Brian knew that if I called for a no-shit defensive turn, it meant electronically I was getting some pretty serious signals, and every nanosecond counted at that point. Even changing our heading by only two degrees could destroy the missiles' solution at the speeds we would be flying.

We talked over our procedures for lost communication. If we lost the ability to talk to each other, we had devised a set of signals to relay between cockpits, via certain warning lights or navigation displays that we would both see. This way we would know whether to continue the mission or abort. If it came down to bailing out, the plane was equipped with warning lights to convey that between cockpits.

Ready Sled.

THE CRITICAL FORCE...

I first came to Lockheed as an associate engineer in structural design. I was put to work on the T-33 and the P2V and later the P3. Sometime in 1959, the Skunk Works put out a request for a young engineer. To my surprise, my boss recommended me for the job. I was surprised because at that time, none of us knew exactly what was going on over there, as the security was very good. I always felt, with that much secrecy, whatever was going on at the Skunk Works would be very interesting to work with. I naturally assumed that they would want only older, more experienced people for the job. Another engineer from my shop had been asked to go over there but he turned it down saying he didn't want to have anything to do with such a secret operation. It intrigued me somewhat. I had only been with Lockheed four years, but in 1960 I was assigned to the Skunk Works.

 Upon arriving at the plant, I was surprised when I did not get a formal interview. Instead, Kelly Johnson simply said that if my boss, whom he knew, said I was right for the job, then he believed him and no interview was necessary. I had no idea exactly what I would be doing but realized it must be very classified due to the amount of security I saw.

Finally I was assigned to a working group for fuselage construction. This was just a few months before Gary Powers was shot down in the U-2 and already quite a bit of preliminary work had been done on a new plane. I was amazed at how small my group was — just six guys. Ed Baldwin was the leader of the group and one day showed me the drawings of the entire aircraft we would be constructing. It was pretty amazing, but what also impressed me was the fact that these guys were making detailed drawings so matter of factly. Kelly truly had some talented people there. As the youngest one there, I often felt a little intimidated, but they treated me with a great deal of professional respect.

Construction at the plant.

IRON NEGOTIATIONS 47

Ed had created an entire tenth scale structural map of the new aircraft. I was very impressed, not only with the plane, but with Ed's detailed drawing. I then knew that whenever Ed disappeared for a while, it meant he was working on some new detailed three-view drawing that Kelly needed to have. These people were marvelous in their talents and were able to turn out quality work in half the time I had been used to seeing it done.

Kelly's office was close to ours and he made himself very visible to his employees. He wanted only a few good people, and then he got out of their way and let them do their work. It was one of the many ingredients which made everyone attain their highest levels of quality work. Kelly would accept mistakes but demanded a quick recovery. He always said that the very worst decision was no decision at all. There were many problems which needed solving and much compromising necessary.

The guys working on some of the plumbing that ran through the fuselage said that they needed to drill a hole in one of our bulkheads. Well, to Ed Baldwin, that was his bulkhead, and no one was going to drill a hole in it unless he thought that's where it should be. In situations like this, I think Kelly Johnson demonstrated his true genius. He had put together a unique group of talented and headstrong engineers who, of course thought their way was best, and then he had the innate ability to make them all work together for a common goal. Ed, Kelly, and the other group leader all headed for Kelly's office and spent about five minutes, first giving their opinions, which Kelly always listened to, and then listening to Kelly speak. Everyone exited calmly and work continued. No hole was drilled in Ed's bulkhead.

The Materials group was fairly large since they had many new problems to solve. They had to develop new types of fasteners since normal rivets would not do with the titanium. No cadmium tools were allowed either since they would weaken the titanium. To take care of some problems quickly, we would nickel plate a bolt to cover the cadmium. No one had ever seen the finished product, so we were all anxious to get it built.

I'll never, ever, forget that first time we completed the jet. The final assembly area was right next to our work area, and when they finally pulled all the work stands away from the plane, there it stood. Here was the realization of those three-view drawings I had seen eight months earlier. The world hadn't even seen it yet and there we were admiring this magnificent creation in a way that I don't ever remember with any other assembly line. They didn't waste any time getting it up to the ranch either, as Kelly was very anxious to meet the flight dates that he had forecast.

As I got more experience, it was finally my turn to go up to the ranch, and even though it was quite cold in December out there, I was excited about the prospect of getting to see this plane fly. I was sent up to help with the modifications to the vertical fin. They were initially straight up and then modified to angle inward. I normally spent so much time in the hangar that I never got to watch the plane take off, but one day I had some time and really enjoyed watching the launch. I felt the sound from a mile away.

There was much anticipation about the flight dates which Kelly had committed us to meet. We knew he rarely backed off on those things, so we had some pretty hectic schedules to meet. Part of the fuselage wasn't quite right for Kelly, and he wanted it fixed prior to the high brass coming out to the ranch to view a flight. Kelly wanted a fillet covering the area where the fuselage met the wing. He wanted it to be removable and to be as light as possible. There was one man he knew he could count on to come up with the fix, and that was a guy named Leroy out of our fuselage group.

Now Leroy was quite an innovator and had quite a reputation for improvising. Kelly was adamant about presenting a good flight for the dignitaries on the scheduled day, so he told Leroy that he didn't care how he did it, just make them lightweight, even if the damn things fell off, just do it.

Well, of course Leroy went right to work with all the enthusiasm of a young schoolboy. The plane was to fly on Monday, and on the previous Friday, Kelly decided it would be a good idea to have a practice flight just to make sure everything was in order. On takeoff, all of Leroy's panels came right off. Leroy about died and was right on the phone to us at the plant to get right up there.

SKUNK WORKS OPERATING RULES
DEVELOPED BY KELLY JOHNSON

1. THE SKUNK WORKS MANAGER MUST BE DELEGATED PRACTICALLY COMPLETE CONTROL OF HIS PROGRAM IN ALL ASPECTS. HE SHOULD REPORT TO A DIVISION PRESIDENT OR HIGHER.
2. STRONG BUT SMALL PROJECT OFFICES MUST BE PROVIDED BOTH BY THE MILITARY AND INDUSTRY.
3. THE NUMBER OF PEOPLE HAVING ANY CONNECTION WITH THE PROJECT MUST BE RESTRICTED IN AN ALMOST VICIOUS MANNER. USE A SMALL NUMBER OF GOOD PEOPLE (10% TO 25% COMPARED TO THE SO-CALLED NORMAL SYSTEMS).
4. A VERY SIMPLE DRAWING AND DRAWING RELEASE SYSTEM WITH GREAT FLEXIBILITY FOR MAKING CHANGES MUST BE PROVIDED.
5. THERE MUST BE A MINIMUM NUMBER OF REPORTS REQUIRED, BUT IMPORTANT WORK MUST BE RECORDED THOROUGHLY.
6. THERE MUST BE A MONTHLY COST REVIEW COVERING NOT ONLY WHAT HAS BEEN SPENT AND COMMITTED BUT ALSO PROJECTED COSTS TO THE CONCLUSION OF THE PROGRAM. DON'T HAVE THE BOOKS NINETY DAYS LATE AND DON'T SURPRISE THE CUSTOMER WITH SUDDEN OVERRUNS.
7. THE CONTRACTOR MUST BE DELEGATED AND MUST ASSUME MORE THAN NORMAL RESPONSIBILITY TO GET GOOD VENDOR BIDS FOR SUBCONTRACT WORK ON THE PROJECT. COMMERCIAL BID PROCEDURES ARE VERY OFTEN BETTER THAN MILITARY ONES.
8. THE INSPECTION SYSTEM CURRENTLY USED BY ADP, SHOULD BE USED. PUSH MORE BASIC INSPECTION BACK TO THE VENDOR – DON'T PAY FOR PIECES THAT DON'T WORK! DON'T DUPLICATE SO MUCH INSPECTION.
9. THE CONTRACTOR MUST BE DELEGATED THE AUTHORITY TO TEST HIS FINAL PRODUCT IN FLIGHT. HE CAN AND MUST TEST IT IN THE INITIAL STAGES. IF HE DOESN'T HE RAPIDLY LOSES HIS COMPETENCY TO DESIGN OTHER VEHICLES.
10. THE SPECIFICATIONS APPLYING TO THE PROJECT MUST BE AGREED TO IN ADVANCE OF CONTRACTING. BE SURE THERE IS MUTUAL UNDERSTANDING IN THIS FIELD BEFORE PROCEEDING; OTHERWISE IT TAKES A MAMMOTH CONTRACTING DEPARTMENT TO UNSCRAMBLE THE MESS THAT NORMALLY DEVELOPS.
11. FUNDING A PROGRAM MUST BE TIMELY SO THAT THE CONTRACTOR DOESN'T HAVE TO KEEP RUNNING TO THE BANK TO SUPPORT GOVERNMENT PROJECTS.
12. THERE MUST BE MUTUAL TRUST BETWEEN THE MILITARY PROJECT ORGANIZATION AND THE CONTRACTOR, WITH VERY CLOSE COOPERATION AND LIAISON ON A DAY TO DAY BASIS. THIS CUTS DOWN MISUNDERSTANDINGS AND CORRESPONDENCE TO AN ABSOLUTE MINIMUM.
13. ACCESS BY OUTSIDERS TO THE PROJECT AND ITS PERSONNEL MUST BE STRICTLY CONTROLLED BY APPROPRIATE SECURITY MEASURES.
14. BECAUSE ONLY A FEW PEOPLE WILL BE USED IN ENGINEERING AND MOST OTHER AREAS, WAYS MUST BE PROVIDED TO REWARD GOOD PERFORMANCE BY PAY NOT BASED ON THE NUMBER OF PERSONNEL SUPERVISED.

Kelly very simply stated, "We are flying this plane Monday. See to it this 'little problem' is fixed." Kelly knew we would eventually come up with a better solution to the problem, but he also understood the importance of showing the aircraft on schedule.

We worked around the clock the entire weekend. Poor Leroy was dead on his feet by Sunday, and we made him get some sleep. The flight went off on Monday without a hitch. Naturally, much later we gave ol' Leroy a great deal of kidding over his infamous falling panels. And of course, in time we did come up with a better design, and that is why you see those little pie-shaped panels on the leading edges of the wing.

For an airplane that size it really had a thin skin, and no one could come up with a fuel tank sealant that could withstand the enormous heat generated in flight. We later got into trying to deal with the problems of fuselage expansion in flight too, and this made it even more difficult to locate the source of leaks.

From the time I started on the project, we did three different fuselage designs. We had the first A model, then the YF-12, and finally the SR-71. The wing and nacelle area never changed, but the fuselage sure did for us. Remember, we went through models of one seat, then two seats, with space for weapons bays, and then the reconnaissance version which had a slightly longer fuselage. Overall, we built thirty-five airframes.

We realized even then that we were on the cutting edge of technology and everyone seemed to really enjoy their work. It seemed that Kelly was able to put together a group of people that all had the right work ethic. We all wanted to do our best for him. He was certainly the driving influence in every phase. And I remember we had fun too, and there were a lot of practical jokes during break times. But when it came time to work, you could hear a pin drop in that place; there was no fooling around. Of course, every morning like clockwork ol' Leroy would have to use the bathroom and every morning, right on schedule, he would sneeze and cough like he was going to gag, as he was allergic to that Borax soap, and this sound echoed across our entire area, every morning. Then he would come out and we would all applaud, every day.

Except for certain repairs and modifications, our job was done when the planes left the factory. All during this time there were people working behind the scenes on other projects for future use. The things we were asked to do in such a short span of time are amazing now, looking back, because we really didn't know how to forge titanium in those days. We did everything from scratch, from fabrications to inventing special machines to extrude the titanium, yet accomplished delivering the finished product so much quicker than could ever be done today.

What is most remarkable to me now, is just how quickly, for us, the program came and went. I came into the plant in 1960 and by 1964 everything we were ever going to do with the construction of this plane was completed. We went from the drawings to completing all the planes that were ever going to be built in that short time span. Quickly, some people went on to work with other projects as that plane continued to make flying history for all those years. I stayed with the "black world" projects.

For me, it was the highlight of my career to be a part of something so special. In later years I was one of the chief engineers with the Stealth Fighter and to be honest, the program was a major step down from what we had seen with the SR-71. Military and budgetary red tape had finally caught up with us and it just wasn't the same type of job satisfaction.

I'm glad I got to be a part of the Blackbird program. It was something we will not see the likes of again. Most unforgettable.

— *Sam Kelder* —
Structural Design Engineer
20 years

". . . Nobody is capable of making decisions anymore. There are too many layers of bureaucracy, both in the companies and in the government. As a result, you can't do anything totally stupid, and you can't do anything totally brilliant."

KELLY JOHNSON

DISTANT THUNDER

FROM THE FRONT SEAT . . .

DAY ONE
1725

    We discuss the mission well into the afternoon and realize that we soon need to get some sleep if we are going to be reporting at 3 A.M. As we return to our rooms, I still find it hard to believe that the '111s are actually going to strike Libya. Somehow, it all seems a bit surreal.

    Walt and I talk in our rooms, mostly about the threat and divert options. Though we are very different people, Walt and I think in concert when it comes to the mission. We both spent time in the tactical fighter community before coming to the recce business, and this meant we both spoke the same language when it came time to discuss tactics, threats and options.

    We were supposed to eat a little dinner and then get some sleep, but that was difficult to do as neither of us seemed sleepy. I turned on the TV, and after an hour of British programming, was more interested in my bed. As I tried to sleep, I kept thinking about the fighter guys just a few miles away, at RAF Lakenheath, who were probably already donning their equipment for the long

KC-135Q on takeoff.

mission ahead. I felt a little nervous for them. So much is unknown about what might happen once actually over Libya.

I am still wide awake when I hear the familiar roar of '111s launching from Lakenheath. Interspersed with the distinctive howl of afterburners is the sound of tankers also taking off. The scope of this mission is staggering when one stops to consider the round-trip distance, and the massive tanker support required.

From the many classified briefings that Walt and I have received since coming to the SR-71 program, I also know that many more units and agencies are involved in this operation in a variety of ways, many of them covert. The jets are all off, and the resuming quiet is deafening, keeping me awake for some time.

FROM THE REAR SEAT . . .

DAY ONE
1530

Once Brian and I had discussed most everything, I got really involved with the rather large communications kit that was handed me for this mission. It contained a myriad of codes and priority message information. There wasn't much Brian could do to help me with this; he just stared at that large book and shook his head and was glad he didn't have to mess with it. It was a bit of a pain but was critical in knowing if the primary jet was a go or no-go. It also gave us all the codes for messages that could abort the entire mission or warn us of new threats. Not too many missions ago, we had been aborted approaching the coast of Cuba, and it was a good thing that I was prepared for it, or it could have been a real goat-rope. If a mission went according to plan, I might only use about one-tenth of the checklist, but one never knew. I would be monitoring a minimum of three different radios at any given time in flight. I mostly conferred with the other RSO concerning the many "what ifs" involved with the comm kit.

The entire mission planning process took several hours, but it went by very quickly as our crew rest time was fast approaching. The first SR-71 was going to take off at 0500, so we would be getting up around 0200. Somehow, we were supposed to go back to our rooms right after mission planning and get some sleep.

Brian and I returned to our rooms and talked about a few things, mostly some of his preferences concerning the divert bases. He liked the idea that I had flown in the Med before and was familiar with some of the bases we might have to use. Going into a foreign field for the first time while having an emergency is no fun. Brian asked me a lot of good questions about certain fields that I had seen in my earlier days in the F-4 and F-111. We did have a few differences of opinion concerning certain options along the route, but we quickly worked them out as we spoke quietly in his room. We sort of had a pact that I never told him how to fly the jet, and he, in turn, would recognize that his liberal arts education could never possibly be a match for the analytical expertise derived from four years of engineering school.

I fixed myself some dinner in my room and tried to relax. I kept thinking that surely, having nearly twenty KC-10s on the airfield must be a real tip to anyone noticing such things. I thought about things like seeing missile contrails coming up at us. I wondered if we would see the detonation if it missed us. I wondered if there would be columns of smoke still rising from the bombing raid. Would we have to go all the way because of a problem with the first SR? How would the weather be? Would the cameras be able to see everything necessary? And speaking of weather, what would the temperatures above 60,000 feet be? If they were significantly warmer than normal, it would have a negative impact on our climb performance and could affect our speed. How would our jet fly? What threats would there really be? I wonder if the squadron back home has filled the wives in. These types of questions skip across my mind like a flat stone across a pond, and keep me from going to sleep. I rest easy in the knowledge that Brian and I are a competent team. We've both seen combat. I'm glad that I was crewed with someone out of the Tactical Air Command. We both think alike when it comes to getting the mission accomplished, and it is one less thing I have to worry about.

Later in the evening, as I was about to fall asleep, I heard the angry roar of F-111 afterburners as the jets took off from Lakenheath. I would hear a series of '111s, then a KC-10. After about forty minutes of deafening jet noise, all was quiet. Sometime around 2000, I fell asleep.

THE CRITICAL FORCE...

I got a glimpse of the jet back in 1966 when I was working on a special project at Eglin Air Force Base. It was going through temperature tests in the large climatic hangar there. I thought it was pretty unbelievable at the time, seeing this huge black airplane sitting there covered with a coat of frost. It was something out of a futuristic painting. Sometime later, before I ever came to the program, I remember seeing it fly overhead while duck hunting in the Marysville, California, area. I immediately recognized that distinctive shape from what I had seen in the hangar back at Eglin. I had no idea the plane was based at Beale during that time. It would be years before I ever came to work on the program but I knew this aircraft intrigued me.

I think for some of us, one of our proudest accomplishments was the setting up of DET 4 in England. It was quite a challenge as the local Brits were constructing our aircraft shelters for us. They were a bit unfamiliar with this type of jet and, of course, the hangar ended up being too small for the plane. It took some really quick in-the-field engineering to fix it all, but we did. It was a strategic location for the mission of this plane and everyone seemed to understand the importance of getting operational as quickly as possible. It could have easily become a logistical nightmare, but some very hard working folks at Beale, in England, and at DET 6, coordinated their efforts in a very productive manner, resulting in an excellent facility for the SR-71 to perform important reconnaissance missions.

The Brits were quite amazed at the size of the engine. When they watched the first few engine runs on the test stand, they thought it was for some sort of rocket. Of course, the fact that we blew the bolted rear panels out the back of the hangar while in full 'burner probably impressed them.

Getting fire departments to believe that leaking fuel was no problem was, well, never fully accomplished. You really could douse a lit match in that JP-7. That was the whole reason for TEB on board. If we had used normal jet fuel on the Blackbird, the crews would have been sitting in a flying bomb, risking in-flight explosion with all the leaking of fuel through the expansion joints.

I remember trying to get a picture of the TEB lighting off during one of our engine runs in the shelter in England. Well, we hadn't done too many runs in that new shelter yet and I don't think everyone was prepared for the results. We called that specific shelter the "hush house" since it was built for indoor engine runs and was designed to muffle the incredible sound of the mighty J-58s. The engines were aligned with large tunnel-like metal tubes in the rear of the hangar. Normally, this type of engine run had been accomplished outside, but the "hush house" in England provided good protection from the bitter weather and good sound suppression so, like most phases of this program, here we were about to learn something about performing engine runs in an enclosed shelter.

As I snapped the picture of the TEB lighting off, I then held my position, close to the back of the hangar and thought it would be interesting to watch the full afterburner lighting. Yes, yes, you could say it was mighty interesting. In all my years with the program, I never remember feeling so terrified. As that big flame shot past me, my entire body was shaking from the vibration, and at first I thought that was the reason my shoes were sliding along the wet floor of the hangar. Then I realized that, like a ball of lint being vacuumed off a tabletop, I was slowly being sucked into the metal tunnel which was swallowing the roaring 'burner flame. Test stands, and other support equipment not secured, began to slowly roll toward the back of the shelter. The vacuum effect created by that engine in full blower was amazing. They shut the engine down and, subsequently, engine run procedures were changed to ensure that the front hangar door was not completely closed during runs. My picture didn't even come out very well.

Somehow, we worked out all the bugs at DET 4 and went on to fly some very important missions from that location. It always amazed me that about 40 civilian maintenance guys were doing, over there, what it took about 150 Air Force people to do back at Beale. These were dedicated folks. They worked some very long hours in some very bitterly cold and wet weather. Seeing all that made me all the more intense upon returning to DET 6, and those guys knew that we would support them in any way we could.

— Bill Umble —
J-58 Engine Program Manager
14 years

Full 'burner in the hush house.

CHAPTER 3

The Eldorado Trail

From the Front Seat...

15 April, 1986
Day Two
0300

When we arrived at the DET, we realized that the actual bombing of Libya had already taken place, but that we wouldn't get any reliable word about any of it until the fighters returned. We began looking over mission materials and Walt quickly buried his head in the communications checklist to go over the intricate details of coded radio procedures for this mission.

I reviewed the route and was concerned with the temperatures at higher altitudes, as this would greatly affect the jet's performance. I had already drawn up my own little map of the entire route which would fit nicely in a plastic page in my knee checklist, and I reviewed it as we waited for the main mission briefing. This gave me a "big picture" view of our entire route and helped greatly with keeping correctly oriented during flight.

During the threat analysis portion of the briefing, we note with great interest that many of the Libyan missile sites thought to have been neutralized by U.S. Navy air strikes several weeks ago are still active. Heavy red circles on the map denote anti-aircraft sites, and the coast of Libya seems to glow bright red as the briefing continues. We have seen this type of threat analysis before, but somehow, following last night's raid it seems more pertinent. We know that Qaddafi will likely use less restraint in firing at us than we have experienced in other scenarios.

0330

Walt and I set up the cockpits for the other crew as they are getting suited up. I go over every switch position to ensure everything is set for engine start. I also double-check some little things which can cause a real headache if not set properly. I note the small TEB counters on the throttle quadrant. When fully serviced, each engine had enough for about sixteen shots of TEB. Each time the engine is started or the afterburner is lit, it takes a shot of TEB. The little counters simply gave the pilot an indication of how many shots he has left. Normally sixteen shots are ample as the 'burners are only lit five or six times in a flight, and hopefully the only engine starts are the ones that occur in the hangar.

The little counters themselves were simply a mechanical device that the crew chief set when the jet had been serviced with TEB. Recently, Walter and I had found out what happens when the TEB doesn't get serviced. Apparently, someone saw the TEB counters in our plane reading around two, and simply reset them to sixteen. Someone else then assumed that the plane had been serviced with a full load of TEB. Somewhere off the northern coast of Norway, I couldn't get the 'burners to light at all for our acceleration climb. After trying everything in the book, and some things that weren't, we drove home subsonically. After landing, we found out there was no TEB in the jet, even though the counters were saying there was. Doing that last 900 miles at 450 knots instead of 1,850 knots really gave us an appreciation of just how far from home we really go.

Besides checking the TEB counters, I ensure the projected map display is loaded with the proper route. There isn't much problem of getting it mixed up with any of the other maps this time, as the Libyan mission is a totally new and original routing. I also check to make sure the liquid nitrogen is being serviced and that the fuel agrees with what the mission summary says. For this mission, the jet will take off slightly heavier than normal since, instead of the usual 46,000 pounds of fuel, the big round gauge is reading 65,000 pounds.

Once I am done with the cockpit, I climb out of the plane and walk back on the stand to see how Walt is doing in the rear cockpit. He always takes longer, as there are more nuances to the equipment in the rear seat. Walter is extremely thorough and I really like that about him. When he is through, I ask Walt for a time hack from the super accurate ANS clock so I can set my watch.

I notice that the ANS unit for the other jet is being hoisted into position in its small cubicle directly behind the second cockpit. I always considered it a sort of third crew member. We never left home without it. When Walt is finished, we wait for the other crew.

Once the first plane was successfully launched, we went to PSD and got into our space suits. We then were driven in the van out to our plane, which sat outside of the hangar in a slight drizzle.

From the Rear Seat . . .

Day Two
0212

Like I had done on every mission before, I awakened just prior to my alarm clock going off. And, as usual, I listened to hear Brian stirring in the adjoining room. If either of us didn't hear the other, we'd bang on the door to ensure no one overslept. I stuck to my normal flight routine and ate a small bland breakfast. This would usually hold me through the first refueling. After that I would rely on the little tube food I would take with me in the cockpit. The duration of this mission really wouldn't be too long for us, especially if we didn't have to relieve the primary jet. We were used to flying longer sorties as the norm out of England.

As I ate my breakfast, I thought of my buddies in the F-111s, and that long twelve hours or more they were going to be in that cockpit. I've done that, and you can only squirm so much in that hard seat after about the first six hours.

0255

As Brian and I walked over to the DET, we noticed our tankers launching into the darkness. They would have a long day.

When we arrived at the squadron, we asked a lot of questions about the night's bombing mission. The first reports indicated heavy losses of U.S. aircraft. We all, of course, suspected these first reports of being erroneous and knew that no accurate data would be received until the jets were back into U.K. airspace.

The DET commander, Colonel Barry MacKean, was a former SR-71 backseater, and he wanted to go over the comm procedures again. We were beginning to make it too complicated and I suggested deleting some of the transmissions to simplify things. One thing I always really admired about Colonel MacKean was that he was a real straight talker, unlike some military commanders. When Barry spoke, you didn't need to decode what was said, he put it right on the line. He agreed with me and made some command decisions that were very helpful to the guys in the back seat that day. It was great working with someone like that, especially in the high priority, high tension environment we were now in.

In the hangar, a jet is readied.

Because both jets were launching, Brian and I would be the mobile crew for the first launch, and then come back in, suit up, and then take off ourselves. This was pretty unusual, as I don't ever remember being the mobile crew and flying in the same day.

As mobile crew, we were to set up the cockpits for the other crew, and assist them from the mobile car via radio in any way they might need. We didn't need to worry about the car too much, as the commander and ops officer would be out there on this one, and the launches were going to be comm out.

The primary jet looked good. I double-checked all the navigational data for the other RSO, to ensure there wouldn't be some error that could cause a delay. The first plane took off without a problem.

Pre-takeoff checks for a wet sled.

THE CRITICAL FORCE...

For a long time, most people didn't know my exact job title, since in those days the name of the contractor was still classified. I actually worked for Northrop who supplied the astro-inertial navigation system for the SR-71. On paper I was the senior field and training engineer. This put me somewhere between the Air Force and Lockheed and normally having to please both.

The guidance group, or what we ended up simply calling the ANS, was not a simple system. It was pretty far ahead of its time. When it was being built, the separate vendors supplying different components, were not told the exact purpose for the finished product. Parameters were very precise, and due to the classified nature of the work, information was not shared between different people on the project. This made for some very precise work with strict tolerances but later produced little in the way of technical manuals that could be used in the field when the system became operational on the SR-71.

I was fairly unique amongst the tech reps, in that I had started out as an Air Force sergeant in aircraft maintenance on the F-4, and then was selected to come to the SR-71 as a staff sergeant working with aircraft flight controls. During that time, the Air Force was picking only its top maintenance people to come to the program, and I had no idea exactly what it was I had been assigned to when I was told to leave Clark Air Base in the Philippines and report to Beale.

I was told nothing about my new job until all sorts of security checks were met, and then I got to "meet" the airplane. It was quite impressive to say the least. The way they guarded that thing, you would have thought there was a bullion of gold in the cockpit all the time.

The Air Force philosophy and general approach to the aircraft was much different in the early days than what I saw later in the program. As a maintenance sergeant, I not only learned my area of flight controls and autopilot but was eventually required to learn about all the other systems on the plane. The thinking was that when a problem arose, work wouldn't have to stop because the specialist wasn't there or there was a shift change. Everyone would know a lot about all the systems. Little did I know at the time how valuable this training would be for me in my civilian role later with the plane.

By the time I got out of the Air Force in 1974, I had an excellent working knowledge of how the various components of the SR-71 interfaced with each other. I went to work for Litton Industries and took some courses in computer technology at the same time. Without my realizing it, everything I had done up to that point had prepared me for my most important work on the program. When the Northrop job for ANS field rep became available, I was approached immediately because of my previous work with the plane. I was very pleased with the idea of going back to that airplane and its mission.

The only way I was able to become intimately familiar with the navigation system was to spend time in the laboratory where guidance group (that's what the system was called) problems were worked. My computer training proved invaluable. Because of the secrecy surrounding this equipment, there was very little printed data to reference. The people who had been in on the original design were long gone, and remember, they didn't even talk to each other, so I had to learn as much as possible about the system and then be able to translate this to the guys working with it in the field.

This type of navigation system was extremely accurate. The same technology had been used in certain missile guidance systems. In the SR-71, the ANS was more than just a navigation system, however. Like a mechanical brain, it also interfaced with activation of onboard sensors and was linked to the autoflight control system. This was accomplished by loading a tape into the ANS which was programmed for one specific route of flight. The system not only knew how to navigate the aircraft precisely along that route but, with the autopilot engaged, knew exactly when to start the aircraft into a turn so as to roll out right on course, using the precise angle of bank required, all the while monitoring time and position so as to activate cameras at preset times during the mission. Doing this accurately at speeds of 2,000 miles per hour and altitudes of 80,000 feet was quite an accomplishment. Needless to say, if this system was not functioning properly, the mission did not go. Subtle errors in the system could have disastrous results, and the RSO was constantly checking the system in flight.

The accuracy of the system was phenomenal. This was largely due to the fact that it was a star-tracking system. Some of this I still do not want to talk about, but I can tell you that it could track stars in broad daylight. It was programmed to fix the position of sixty-one different stars. If the ANS could locate just two stars, it could fix its position, although three stars were better.

Normally, the navigation was, well, dead on. We considered a tenth of a mile deviation about the maximum error we were comfortable with. For over twenty years the flying crews were quite comfortable with the ANS, a system which stayed with the jet for the life of the program.

There were so many components to that system, you really had to have faith in computer technology to believe it was going to all work properly, especially when putting it all into the environment of high Mach flight. First, of course, was the intense heat, not only inside the plane but across the skin externally. The face of the star tracker lens sat right behind the RSO's cockpit and it was programmed to compensate for the heating and minute warping of the fuselage in flight. The computer also compensated for shock wave disturbance. Keep in mind, too, that the earth is not standing still beneath the jet, so its relative movement was also programmed into the system.

The accuracy of the ANS used to really impress the new crews. When the jet was holding short of the runway prior to departure, the RSOs would run a check on the system to see if the ANS knew where it was. Most regular inertial systems would simply say that the jet was on the ramp at Beale, for example, heading 210 degrees, going 0 knots. Our system was more on the order of telling the crew that they were on the north end of the field at Beale, 500 feet from the runway, in parking spot bravo, heading 210, going 0.00 knots, but there was a 10 knot wind blowing across the jet, so check with the pilot to make sure you are not rolling at 10 knots.

Naturally, fixing one's position this accurately requires an accurate timing device. Well, the system actually incorporated an atomic clock. This little gem was accurate to one part in 10^8. Before inserting the system into the aircraft, we would set the clock in the shop and even factor in the delay, in nanoseconds, of the signal from the transmitter clock to the receiver and take into account signal phasing. Few people realized it, but each one of those atomic clock crystals were tracked by the Nuclear Regulatory Agency, so we had to account for all of them all the time. Needless to say, we all used to set our watches by the time in the shop.

In the central computer of the ANS, we started out with only 32K computer RAM memory. This was a remarkably small amount considering the numerous tasks assigned to the ANS. We eventually installed 65K of memory. When the aircraft was upgraded with DAFICS, we had to upgrade our software but remained at 65K memory. I will say we performed some magic with that software to make it all work for a successful mission.

When I returned to Beale as a tech rep, about ten of the ANS units were down for maintenance, so I had a chance firsthand to learn everything about keeping this system working. My background in maintenance proved to be invaluable. There was so little documentation for the system at that time that we began to write the book on how to keep this thing operational. We felt, next to the engines, our system was the most important for mission accomplishment. I spent most of my time training Air Force people in the maintenance of the unit, and of course interfaced often with the other shops, to include sensors, autopilot and flight instruments.

I saw the Air Force go through many different managerial philosophies over the years, and unfortunately for the program, there gradually came the idea that tech reps were expendable and the Air Force should be able to handle all aspects of operating the SR-71. This was a mistake, as long as the Air Force continued the policy of moving people around and bringing new inexperienced folks into the program. While managerial philosophies were changing, the delicacy, specialness and complexity of the SR-71 was not.

As a tech rep I was a valuable link to each new group of people for training. It was the type of program where corporate knowledge meant a great deal, and basically we had it. Near the end of the program, I knew more about that ANS and how to keep it operational than anyone else, and I also knew what errors we were making in how we managed the system. At a time when military budget cuts were making most of the headlines, the Air Force was increasingly less interested in proposals to rewrite tech data and fix some inadequacies in the system. I think I worked the hardest

Close view of cockpit windows, circular ANS window, and refueling receptacle.

near the end of the program, ensuring that the ANS maintained the excellence it had always provided. In the end, I was faced with dealing with less experienced people coming to the program and increasingly less support from the Air Force as a whole.

When it was all over and we found ourselves out of a job, I was very proud looking back that I was a part of an elite team that performed an important mission. When we would see the crews return, sweat pouring from their faces, and exhausted, and we knew we got all the data, it was a proud feeling to be working with such professionals.

The ANS units are all stored away, somewhere, now. A few may currently be used on some missile tests. That was quite a system, well ahead of its time, and a great match with the Blackbird.

— *Jack Levine* —
Tech Rep, Astro-Inertial Navigation System
13 years

"[The tech rep] . . . *helps maintenance personnel trace and identify the source of integral fuel leaks, and assists with procedures involved in repair and resealing of fuel tank system. Assists fuel shop in troubleshooting and repair of SR-71 fuel systems, including air refueling latch, fuel dump, fuel vent, the liquid nitrogen system, and fuel cooling. Trains fuel shop people on MRS data evaluations. Provides assistance on drag chute system to include jettison and release mechanism. Assists tire, wheel and brake maintenance shop with any problems. Assists shops performing any maintenance on canopy or ejection seats. Assists in the maintenance and repair of ground support equipment. Assists with servicing of the aircraft. Trains new personnel on proper aircraft servicing and aircraft lubrication. Maintains reference files of technical manuals and blueprints.*

FROM THE OFFICIAL JOB DESCRIPTION OF TYPICAL TECH REP AT BEALE

BY DAWN'S EARLY LIGHT

FROM THE FRONT SEAT...

DAY TWO
0530

As we sit in our space suits in the van, we know we have about ten minutes before we have to get into the jet to make our takeoff time, an hour behind the primary jet. After about two minutes of waiting, Walt and I both decide that we'd rather be sitting in the jet, doing something instead of just waiting and thinking.

It is dark, cold, and a light rain is falling. It is England.

Departing from our normal procedure of starting the planes in the hangar, our jet is parked outside, situated for quick access to the runway. This was so we would deconflict with the large number of tanker aircraft also utilizing the taxiways. If one of the tankers broke and blocked the taxiway, we still had a clear shot to the runway from this position.

Before getting into the plane I pat her nose gently, as I always do. She looks more serious this day.

Once in the plane, I am attended to by several PSD technicians who ensure that every fastener and connection with the suit are checked and double-checked. The oxygen system is checked. It is then turned off and the backup oxygen system is checked. My suit's ability to pressurize is checked and I feel the reassuring increase of air pressure across my arms and legs as the suit begins to inflate. They double-check that my vent hose is secure and then give me a hearty thumbs up and depart.

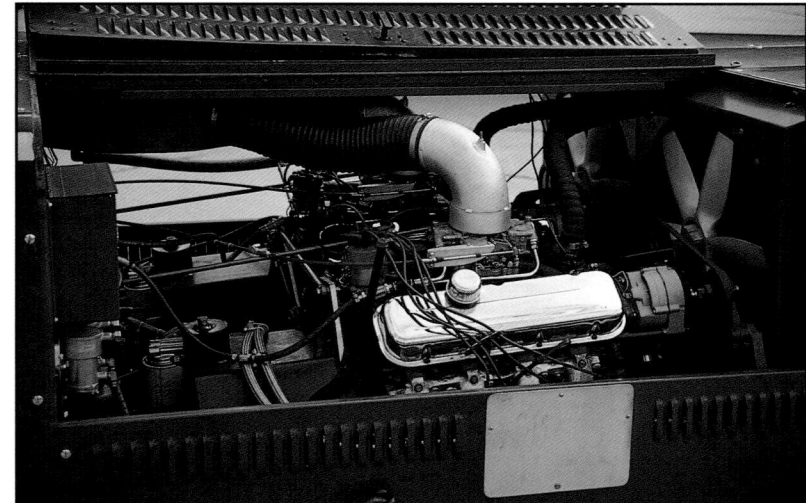

Big block in the start cart.

Once Walt and I have our cockpits set up, we have about fifteen minutes to sit and wait for our engine start time. Since we won't be using the pressurized air start system in the hangar to help crank our engines now, they have rolled out the start cart housing the big block Chevy 454s. We haven't used these in a while for start-up, but I can hear them revving up from where I sit, and I like the familiar sound of those engines.

As the first hint of dawn's light bumps against the dark layer of clouds, I notice the drops of rain that have collected across my clear face plate. I try to wipe them away with my glove, but this doesn't work, and before I can signal to the PSD van for a cloth, there is a technician kneeling on the stand next to me wiping my visor. As the light rain continues, he stays there and continues to keep my visor clear of drops every few minutes. It was very comforting to have these kinds of dedicated professionals on our team.

Finally it is time to start. The Chevy engines wind up to red-line rpm, there is a muffled explosion of TEB, and from the vibrations coming across the rudder pedals and up through my boots, I can feel the very pulse of the jet as it comes alive in a solid purr of idle power.

While running some self-tests with the DAFICS, I have a few moments to think about the individual quirks of this particular jet. Each SR-71 seemed to have a personality all its own. We were in tail number 960, a plane we had already flown a couple times on this tour, so she and I were well acquainted. The last time in 960, we had an autopilot problem and some fairly loose inlet doors.

Heavyweight takeoff at Mildenhall.

I knew our maintenance crews had worked out those problems, but this plane hadn't flown in over ten days, and this was the first flight since the corrective maintenance had been accomplished. I wanted her to perform well today and was relieved to see all the pre-takeoff checks go routinely.

Once on the runway, large chocks were placed against the main tires, and I began the engine runs. As I pushed, one throttle at a time, smoothly up to 100 percent rpm, I watched the familiar dance of the exhaust gas temperature gauge, spinning the numbers up toward 800°C. As the little drum counter spun the numbers up furiously, it always reminded me of a small slot machine. The prize was to observe the engine temperatures stabilize in the nominal range. With the second engine, the temperature readings were very erratic, causing me to take control of the EGT manually, something I wasn't able to do in other jets. With careful bumping of a tiny lever, I could bring the EGT into the desired range. It wasn't an abort item but meant I would have to monitor that engine temp more closely as in manual control there was no auto-temp damping. Jet 960 was not starting out the day impressing me.

With no radio call, the tower flashed a green light toward me, and I knew we were cleared for takeoff. After rolling about 4,500 feet, I started the stick back at 180 knots and expected to feel the nosewheel start to come off the ground as usual but nothing happened. At 190 I continued pulling and still nothing happens. A hundred things race through my mind in the space of a millisecond. Do we have a control problem? Would we have to abort? Then I remember we have 20,000 pounds more fuel than normal as the nosewheel slowly starts up at 200 knots. We are such creatures of habit. Heavyweight takeoffs were rare during normal operations, and every ounce of my body was conditioned to feel that plane lifting off when it wasn't quite ready. She was humbling me early this day.

From the Rear Seat . . .

Day Two
0455

Brian and I went to get suited-up for our flight, and as we always did, we stopped in the men's room for our final pre-flight relief. Brian and I never wore the UCD. It was just too uncomfortable, so that last stop in the bathroom was an important one. Holding it for an entire flight was usually a matter of one's mind over one's body. Occasionally there was an "accident" inside someone's suit, but not often. The suits were airtight, so whatever "deliveries" were made after suiting up were yours for the duration of the flight. Knowing that did help one's resolve at times, but there were some hurried undressings following long missions, and an occasional sprint to the bathroom.

As we came into the PSD area to get suited up, we did everything just the same way we always had. If anyone wanted to see superstitious habit pattern behavior, the suit-up area was the place.

Each crew member had his own specific way of doing things. Brian had this thing about the way he put the long johns on; they had to go on just right or he would be uncomfortable the whole day. I always had to put an even number of articles in the front leg pockets of the space suit. I always wanted a large plastic bottle of orange Gatorade. Brian had to have a small bottle of half water, half green Gatorade. In the nearly five years that Brian and I were together, I don't think we changed one little thing in the suit-up ritual.

One thing aviators try to do is develop useful habits and then stick to them so that in times of stress (frequent in this business) they won't forget something that might save them. I always took with me a backup hack watch set to Greenwich Mean Time from the chronometer in the ANS shop. Once in the jet, I would check to see that my watch and the clock in the jet agreed. I had once heard this horror story of an RSO flying with the atomic clock in the back seat set to the wrong time.

There is no better way to get lost than to have your navigational equipment slaved to the wrong time. Even a minute or two, at 2,100 miles per hour, will make a big difference. I had never seen an error like that in all my time in the plane and of course attributed it to the fact that I carried that little backup watch, so it became a necessary part of my equipment.

The PSD crews were always great but even went out of their way more than usual to ensure we were comfortable for this mission. They, like everyone at the DET that day, seemed to sense the

An RSO leaves PSD van for jet.

increased importance of the situation. I always enjoyed talking with the PSD crews while suiting up. They were always such a professional and completely helpful group of men and women.

When we got to the jet, our minds were filled with each of our individual thoughts, and as usual, Brian and I talked little, shook hands at the bottom of the ladder, and got down to the business of getting settled into our cockpits.

Everything was checking out just a little too perfectly in the back. That made me a little nervous. Then I saw something in the ANS that didn't quite make sense. It was showing an error, and that was pretty unusual at this point. I had spent a lot of time with Jack Levine back at Beale going over the fine points of the ANS, and something he had told me months ago rang clear in my ear now. He said if the ANS was in error before engine start, check to see if the chronometer was off, because if that was the problem, it would be a quicker fix than having to swap out the entire ANS unit. It was the first time I had ever seen my little hack watch actually disagree with the ANS. I was seeing an error of two minutes. That meant the jet would be off course by as many miles as we could fly in two minutes. Back during my days in the C-130, that would mean only nine to twelve miles and we could compensate in other ways. In this jet it meant, while travelling Mach 3, we would be sixty miles off course.

When I alerted the ANS tech rep on the scene as to this discrepancy, he initially didn't think their clock was off, since that was pretty rare. I asked him to check it anyway, and sure enough I was right. At that point, I had to shut the ANS down and re-initialize it, set with the correct time. Brian hated when I told him about these sorts of things. There wasn't a thing he could do to help and he barely understood all that I was trying to tell him about a piece of equipment he never got to use; but he did know it could cause an abort or delay, and he hated that. In a short amount of time, I was able to get everything working properly.

Since we were operating in a "comm-out" mode, there were no transmissions made on the aircraft's radios, and the signal to start was given visually by the commander in the mobile car.

Due to the relatively long distance we must fly to meet our first tanker, we take off heavy with extra fuel, but from my seat I notice nothing different in our departure.

THE CRITICAL FORCE...

I started working with Lockheed in their missile division. With a degree in aeronautical engineering, I came to the Blackbird program in 1964 when they were flight testing the first three jets. Basically I served as a liaison at Edwards AFB between the flight test guys and the engineers back at the plant. Previously, I had experience with test programs on the B-58 and had even flown in that plane, as an engineer observer, at Mach 2. Lockheed sent me to Beale in 1966 as a field rep, and when I arrived, there weren't even any SR-71s yet.

Once we got our first few planes, it began an era of unprecedented operational success that remains to this day the crowning achievement of my varied career as an engineer.

I initially was involved with developing the start carts. It took a lot to crank that J-58 over, and the original start system used on the YF-12 was a pretty wild pyrotechnic sort of device. Starting the engine back then was a sequence of small charge devices firing off, followed by a large cloud of smoke. We later went to the Buick start cart, which we kept for quite a few years. That was a unit with two Buick Wildcat engines in tandem, which provided torque for a shaft to get that engine turning. That was quite a sound in the hangar when those Buicks wound up. I was never totally comfortable with that system though, as we were frequently throwing a rod or burning up one of the Buick engines. We basically had to red-line those V-8s to get enough juice out of them, and sometimes they would fail just short of the rpm which the J-58 required for start. Occasionally we'd see the young troops out on the line, revving those start carts up just for fun, but by and large, they tinkered with and nursed those start carts like they were their own hot rods.

Eventually we had blown up more Buick engine blocks than we could easily replace, especially in those days when American cars were coming out with smaller engines. We were a pretty

Start cart controls.

resourceful group though, and many an auto wreckage yard across the country was contacted at various times to keep us supplied with those big blocks. I'm sure they never realized their contribution to the Blackbird program.

Finally we were forced to go to the bigger Chevy 454. Those things were practically bullet proof, but some of the same problems remained. After years of an impressive sounding, semi-reliable start system, I eventually helped transition the program over to a much more reliable and safer compressed air start system, albeit much less impressive sounding.

I went with the third team over to Kadena. We had a lot to learn, but everyone was so eager to make the program a success. One enthused young airman, a fuel specialist, almost lost his life one day during a routine inspection in the hangar. We were having continued fuel problems with one of the jets, and when it returned from a mission, maintenance was eager to check it out. This one airmen quickly removed the necessary panel and began climbing up into one of the fuselage entrance spaces, called a "boyhole." This was a name given to these bulkhead spaces by the guys back at the Skunk Works, since they were much smaller than manholes. Well, this poor guy gets himself wedged in this boyhole, begins his inspection, and then in a few minutes his whole body goes limp. I just happened to notice him like that and quickly rushed over to help him. Up to that time, we hadn't really thought about the fact that there was no oxygen in those fuel tanks after landing; he had simply passed out from lack of air. We tried to pull him down but couldn't squeeze

Running it up at Kadena Air Base.

him out of there easily. Quickly, someone grabbed an air hose and shoved it up that hole and very quickly he regained consciousness. Of course, the first thing he wanted to do was continue with his inspection. We talked him out of there and had him sent to the hospital, where they found him to be just fine.

While at Kadena, I witnessed the burial of one of the jets too. Sounds unbelievable now, but in essence, that's what we had to do. One of the planes was beyond repair after crash landing on the runway. After we had cannibalized all the parts that we could off of it, we were left with a large chunk of titanium, strongly resembling an SR-71. The decision was made to cut the plane into pieces and sink it far out in the depths of the South China Sea. We couldn't do that, though, since just days before, our country had signed a new international treaty that prohibited dumping materials into the world's oceans, so a new plan was devised. We hauled the pieces out to a nearby hill and had the fire department come out to ignite them so they would be burned beyond recognition.

Well they poured jet fuel all over the aircraft pieces and had a terrific fire burning for quite a while. When it subsided, we found that all the fire had done was burn the paint off and cause a few minor wrinkles in the skin. The basic structure of the plane was still intact. Someone pointed out to the fire department that this plane probably experienced greater temperatures in flight than it had in their fire. We were learning.

JP-7 streaks fuselage on low pass.

They finally gave up on the fire idea and we simply buried the pieces in the ground, on what was forever known as HABU Hill.

I always felt there was a lucky star over this program because somehow we were able to escape unharmed from some very dangerous situations. The fuel which leaked from the plane normally covered the hangar floor and was swept down drains by maintenance crews daily. One morning, we launched a plane, and due to a problem on takeoff roll, the crew aborted, and before anyone could do anything about the fuel on the hangar floor, here came that jet right back in. Even at the slow speed which the pilot used to taxi into the hangar, the jet just kept sliding slowly past us when he applied the brakes. It was then that we became very aware of the tug vehicle which was parked right outside the other end of the open-ended hangar. We felt like we were watching an accident happening in slow motion. That jet just kept sliding across that slippery fuel. We all felt so helpless watching it. One sergeant actually shoved on the nose of the plane in a futile attempt to turn it.

Finally when the nosewheel reached the pavement outside the hangar, the pilot was able to use the nosewheel steering to avoid the tug, but he still didn't have any brakes as his wheels were still thirty feet behind him in the hangar. That plane rolled through the entire hangar and came out the other end without a single scratch. When it was finally stopped we all began breathing again.

The program, though very secret to the public, was very "high visibility" to certain high ranking military people. Every little thing which happened with this plane had to be documented, and often Air Force reports could give the reader a false sense of what was really happening. This is where we were really lucky to have a man like Paul Mellinger on our team. Sometimes I could call him from Kadena and give him the real scoop, enabling him to better put commanders back home at ease over certain developments. I don't know how he did it; I certainly would not have wanted his job. He was the real glue that held things together in this program.

Over the years I went on some interesting trips. Once I was dispatched to Grand Forks AFB to help recover one of our jets. An SR-71 had made an emergency landing on a very icy runway, and in the process, had blown all six main tires. This didn't make the local commander very happy since his runway was now closed, and he proposed towing the plane off as it was, blown tires and all. This would, of course, have caused a great deal of damage to the wheel structure, so we assured him that we could get a team up there, change all the tires and have the runway cleared in a day.

Our airmen worked all night in frigid weather, jacked that plane up, changed all six tires and brake assemblies and had the runway cleared the following day. The commander there assured us that if a SAC alert had commenced with our SR-71 on the runway, they would have bulldozed it off to get his bombers airborne, but he was glad that he didn't have to give that order.

Getting bases which our planes diverted into to believe that the leaking fuel wasn't a problem was another dilemma. It was not unusual for us to have to drain the entire fuel system before certain bases would let us use one of their hangars. No matter how much we briefed local fire departments about the low flammability of our fuel, leaking gas simply made them nervous.

One time, at Kadena, we actually used the aircraft fuel to put out a fire. The old start carts occasionally caused small fires. The regular gas the Buick engines used would sometimes pool inside the cart, then when the carts were pushed toward the jet, the fuel would spill across the hot exhaust pipes and cause a fire on the hangar floor. That very thing happened, and the fire slowly spread across the floor toward the spare jet, and I had visions of the entire hangar going up in flames. Luckily, one Air Force crew chief acted promptly and swept puddled JP-7 toward the fire. The fuel extinguished the fire quickly.

It was a treat to work with a program where people could really get things done. Kelly Johnson was not one for wasting time. We were never asked to do something simply for the sake of satisfying someone's idea of a requirement. We did things which helped get the mission done. It sounds so simple now, but I can tell you that many programs do not work this way today. They are drowning in a sea of red tape and ever-changing requirements lists. Kelly's basic philosophy was to do it right the first time. That way you wouldn't need an army of inspectors later, to show you where you did it wrong. Kelly often said, "If my engineers can't design quality into this jet, I'm going to get rid of them."

Our biggest headache in the field was getting the many people involved with the program to understand Lockheed's position in the program. Originally, the Air Force management team at Wright-Patterson AFB was fairly small, and that worked well. Later, when it got much bigger, too many people were in the loop who weren't as knowledgeable about the program as they needed to be. It was frustrating to watch the process of ever increasing military inspection teams, require-

Tech reps on location.

THE ELDORADO TRAIL 75

ment committees, program managers, etc. Where we used to coordinate with one person, we now had twenty, and that one person knew more about program specifics than the whole twenty.

I doubt that many people are very interested today in all that we did with helping to keep this plane flying. But we were extremely proud to have a part in it all, and we truly felt like pioneers. We were doing things with that plane and running the program in ways that are simply astounding when I think about it today. Most of the civilian people I got to work with had strong ideas about loyalty and dedication. We were from a different generation that I guess was a little unlike what you see today. I worked with some very outstanding Air Force people too. Commanders like Doug Nelson, Pat Halloran, and Bill Pugh, to name a few, were not only great commanders but wonderful people as well.

To me, though, and I think you could find many others that would agree, Paul Mellinger was the single individual most responsible for the operational success of this aircraft. His ability to work with both the military and civilian sides of the program was nothing less than brilliant. Both sides needed a man like Paul, but in the end I don't think either side fully appreciated what he meant to the program. He truly was "Mr. Lockheed."

It is so strange to see the plane in museums now. I hated the way they cut some of the planes up to ship them to some pedestal somewhere. It just didn't seem right. We always took such care with the plane. There was nothing else like it.

— *Jim Cook* —
Tech Rep
20 years

" *Be Quick, Be Quiet, Be Right. Especially Right.* "
                        A SIGN POSTED BY KELLY JOHNSON IN THE WORK AREA OF THE SKUNK WORKS

MORNING PASSAGE

From the Front Seat...

Day Two
0655

On the way out to our tankers, I was able to play with the EGT control and got it to work back in the auto-mode, and that made me feel better, since the engines would be better matched that way. This was the kind of thing that frequently happened when planes sat on the ground for an extended period of time.

As we were cruising subsonically, we received a radio call for conflicting traffic. I immediately picked up the traffic, and to my surprise, it turned out to be the returning F-111s and their tankers. They are only a few hundred feet below our altitude, and slightly offset to our left. We are all above the lower layer of clouds now, and it is a magnificent sight as the low morning sun sharply illuminates the formations of tankers and fighters. Like little chicks around the mother hen, each tanker has a separate set of fighters in tow. It is an impressive line of air power that passes opposite to our heading. I know where those fighters have been and think to myself it must have been one hell of a night for those guys. They've been in the jet over ten hours now and I know they must feel elated to finally see the coast of England. From our experience, Walt and I know that it would be phenomenal for there to be no losses in a mission of this scope, and we both try to count the jets. We think perhaps one or two are missing. We know most of these guys. We are with them in spirit. As we pass the last group in their formation, I rock my wings in salute to them. The '111 at the end of the line, rocks his wings in return.

I spotted our tankers easily as they pulled bright contrails across the early morning sky. Once on the boom, the tanker copilot informs us that their autopilot is out, so they will be hand flying the refueling. From the cockpit of the tanker, the difference between hand flying and having the autopilot on probably seems minimal. Hanging onto the end of a rising and falling boom in the midst of several turns, I felt the difference was considerable. Turbulence on the refueling track does not help. After more of a workout than desired on the first hookup, we departed the tanker and headed south.

Heading toward our turn point into the Med, I run the jet up to speeds that we'll need in the target area, above the Mach 3 which this straight leg calls for. The left forward inlet door is vibrating but not causing any unstarts, so I watch it closely. I am concerned with the vibration, as it could affect the quality of the photos we came to get. Once in the "take" area, my main job is

Precontact position.

" If you dropped a medium sized bomb at Mach 3, with no explosive device installed, it would hit from free-fall with the kinetic energy of 355 million foot pounds. It would penetrate 31 feet of reinforced concrete, or 300 feet of earth."

KELLY JOHNSON

to keep the aircraft flying in a stable manner as Walter works his magic in the back. I didn't like all the little aches and pains that 960 seemed to be displaying from its downtime.

Once we get to the Med, we receive word that the primary SR-71 is a go, so we know after taking on some fuel we'll be turning around and heading home. Subconsciously some of the mission's pressure is relieved for us, and I take a moment to leave the various "ills" of the jet in the cockpit, and look outside.

Slicing through the Strait of Gibraltar is a beautiful sight as the weather is extremely clear. This is all virgin geography to me since I have never flown in this part of the world before. The Mediterranean is a beautiful blue, and islands stand out clearly like small stones in a big lake. The beautiful scene of colorful geography which now spreads before me, stands in stark contrast to the events of the past twenty-four hours which had brought us, in this jet, to this place. From my vantage point, the world looked peacefully in order, and it was difficult to believe that while everyone was going on about their daily business, we were participating in our own private war with one segment of the broad expanse of land to my right. I held no compassion for Qaddafi, but viewing the Mediterranean in all of its glory made our mission momentarily seem uncombat-like. Any war, regardless of its reasons, is a sad thing, and the pristine view from my seat simply reminded me of this fact.

We return to Mildenhall to find the field covered in rain showers. After landing, we are more exhausted than normal from a mission of this relatively short length. I realize that it has a lot to do with the few hours of sleep we've acquired since yesterday.

Pilot's view of tanker.

From the Rear Seat...

DAY TWO
0620

After takeoff, our direction of flight was completely different from our normal departures flown from Mildenhall. Instead of our usual northeasterly heading, we were now heading southwesterly, out toward Land's End. The nice part about this routing for me was that I now had about thirty minutes to work the ranging with the tankers. Our normal departures out of Mildenhall found us hooking up with the tanker after only fifteen minutes of flight. This required somewhat of a juggling act with the radios, for the RSO, in order to handle the departure, the hand-off to the British air controllers, and effect the rendezvous with the tankers. With enough practice, it had become routine. Normally, I'd have Brian preset his radios so that I could switch between his and mine and keep the frequency changing to a minimum.

Sprint to Portugal. View from 77,000 feet.

Flying subsonically across the English landscape for a while gave me some time to run some other checks with the DEF and the cameras. As far as anyone knew, we were just another bit of routine traffic in the early morning sunrise.

We were both anxious to see how the jet would be flying too, but we really wouldn't be able to tell until after the AR when we took it upstairs. The SR-71 definitely had two distinct personalities, subsonic and supersonic. The first was quite predictable and stable; the latter was always a bit of an adventure.

About halfway to our tankers, British radar gave us a call for conflicting traffic on our nose, slightly below us. Brian picked up the traffic visually, and to our surprise, it turned out to be the returning strike force of F-111s accompanied by several KC-10s. We passed quite close to them and Brian rocked the wings. We both tried counting the jets and agreed that at least one or two were missing, but we were glad to see such a large number returning. We knew from experience that one or two of them might have diverted somewhere along the route with a problem. We hoped that was the case.

Cruising subsonically.

Our first refueling track was not the usual straight-line track but, rather, a small racetrack pattern in the sky. The tight turns only added more strain on the receiver pilot, but Brian was an experienced guy and even though we hadn't seen this type of track in many refuelings, he handled it fine. There wasn't much I could do to help with the refueling. Mostly I monitored the fuel to ensure that we were hitting the end point of the track with a full load of gas. Watching our fuel along with the number of minutes left on the track, I could coordinate with the tanker to adjust the flow of gas we were receiving so that full tanks coincided with the end of the track.

I tried not to talk too much to Brian during the refueling as he was busy enough. Usually, this was a time when I could take stock of my cockpit and even have a quick tube food pudding. Of course, just about the time I got that tube out, the tanker navigator would be asking for an update on airspeed, the winds, and our present position. Our ANS was so much more accurate than anything else flying, that the tanker navs liked getting a good check from us for their own corrections. I never minded this since it also helped ensure that these guys would be where they were supposed to be for our return.

Coming off the tanker, we accelerated normally and began a speed run toward the coast of Portugal. The jet seemed all right, but there was a slight buzz from one of the forward doors on the inlet. Nothing too serious, but it could possibly cost us some precious fuel at higher speeds. The mission planners figured everything very precisely for our flights and rarely got us home with any extra fuel.

In the process of figuring range, speed and fuel remaining, they had to insert a value for the air temperatures — an important part of the fuel equation. Colder than normal temps meant better range and some fuel to spare, while warmer temps cost us on performance and fuel. Since forecasting atmospheric conditions is an imprecise science at best, we were constantly checking the outside air temps for ourself in the climb.

As we got to our comm checkpoint, abeam the southern coast of Spain, we received the message that the primary SR-71 was able to make its run over the target area, so we knew we would be turning back after another refueling.

I took a rare peek out my window and the weather was excellent, making for a great view in our turn. We sped home anxious to hear what the primary crew had experienced over Libya.

Ready to remove chocks on taxiway.

THE CRITICAL FORCE...

In 1962 I was working with flight control systems with the German F-104 program at Edwards AFB. I had an idea that some sort of secret program was in the works when I would see certain neighbors of mine go away all week, probably to some secret work site, and then return on the weekends.

This was an exciting time to be at Edwards since there were numerous flight test programs in progress. I ended up doing some interesting work with a modified F-101 which incorporated a side stick controller. This was interesting technology that years later spawned such systems as the F-16 side stick digital flight control system. I also got to do some flight control work with the X-15. In 1964, I was at Edwards when Lyndon Johnson made public through his speech the existence of

Specialists were quick to respond when there was a problem.
Pilot's face says it's probably broke.

the SR-71. Naturally, everyone was quite impressed, though I wasn't quite as surprised as the rest of the public.

Honeywell had an opening for a field rep position with the SR-71 program and I was fortunate enough to get it. I went to Beale for initial training in the flight control system of the plane and found out that they needed to fill a position in Okinawa, where the plane would be based in addition to Beale. I told them if I could take my wife along, then I was their man. Honeywell said that would be fine, but the tour would last six months. My children were all grown and on their own, so my wife and I thought it would be an interesting opportunity to travel, and we decided to go. We left for Okinawa in 1968 with just a few suitcases and a large trunk. We returned nineteen years later in 1987.

For the first seven years at Kadena Air Base, we lived in a one room apartment and there were very few company perks. But it was a great adventure for us, and we made the most of it. When the company asked us if we would stay on, we said yes, and eventually our living conditions improved quite a bit.

The Vietnam War was our number one priority with the jet during those first few years in Okinawa. We filled all of the shelters with a full complement of four SR-71s. The work was intense and with all the flights that we flew over North Vietnam, the atmosphere was definitely one of generating combat sorties.

Unlike some of the other contractors, I was the only field rep for my particular area. I had about nine Air Force people assigned to the flight control shop. We handled the stability augmentation system (SAS) and the autopilot as well. Being the only rep in my shop, of course meant some long days. My wife didn't see a lot of me those first few years. And of course she wasn't allowed to really know much about what I was working with. She would see the plane take off and fly over the base sometimes and decided that the X-15 was no longer her favorite airplane.

Before we ever upgraded to DAFICS in the early '80s, we had to deal with the old analog system. It really gave us a lot of headaches. The early systems used fuel as a coolant to keep the gyros from overheating. There were a lot of small leaks that were not only hard to find, but hard to plug, without disturbing something else. We learned a lot of little tricks about how to keep that SAS working, and it was my job to help train the new guys on all those little tricks.

The original Air Force people were top notch, but as time went by, the newer people were less and less experienced. My corporate knowledge became indispensable. Honeywell was very pleased with the job I was doing and glad that I liked Okinawa since they didn't have any other qualified volunteers for the job. They increased my pay somewhat, but I think I would have stayed anyway. This was the kind of work that really meant something. Every day was an important mission.

The Okinawans called the plane the "habu" after a poisonous black snake indigenous to the island, and they really made a big deal out of it. For us, it was not a glamour thing; we were at war.

Longtime island resident Chuck Wiethoff.

The ever-present mechanic.

It was deadly serious business. Many times our planes were shot at during flights over North Vietnam, and we always waited impatiently for the birds to return safely.

One of the things that was very satisfying to me was being there when the crews returned from a long mission and seeing that certain look on their faces. It was a mixture of fatigue and genuine happiness to have returned to be amongst us again. Their words would speak of other things, but their faces couldn't lie. I instilled in every GI in my shop the idea that the crew was entitled to a 100 percent working aircraft every time they walked up that ladder.

The more time I was there, the more familiar I became with other systems and this helped greatly when troubleshooting certain problems. For a time we were having some difficulty with the rudder servos giving uncommanded inputs. Watching the rudders vibrating while the plane was in the runup area one day, I advised them to shut one engine down. To the commander's credit, he had faith in my judgement and had them shut down the left engine. With one engine running, the rudders completely froze up, validating my suspicions. That's when we instituted the procedure of starting alternate engines each time the plane was cranked up. At that time, the Air Force actually had no requirement for the servo to work on one hydraulic system. I made sure that was changed.

Interestingly, some of the DET commanders I worked with had first come to Kadena as young SR-71 crew members in years previous. I enjoyed seeing their progression through the ranks and I

Rudder check prior to shutdown.

guess I became somewhat of a father figure to some of them. I was always treated with respect by the Air Force commanders, and I greatly appreciated that.

I was lucky in that I truly enjoyed my work. The hardest part was that I couldn't tell my relatives and friends exactly what I was doing that required me to be out of the country for so long. Only my son knew that we had come to Okinawa, but he thought I was working with the B-52s. Our friends pretty much gave up on trying to figure it out. When asked what I did for a living, the standard answer was that I was with Midwest Engineering. Of course there was no such company, but I was definitely not allowed to say I was with Honeywell. My wife and I did get back to the States to visit a few times over the years.

Honeywell pretty much left me alone over there and I only remember a company representative visiting me twice in nineteen years. I didn't mind that at all since overall they treated me pretty well and I had no complaints. I was very happy to be doing the work I was, with the world's premier airplane.

When we finally made the transition to DAFICS, it was much better, because now the system's self-test told us everything we wanted to know about how the gyros and servos were behaving before the plane ever took off. It wasn't without its growing pains, but it definitely was an improvement in the safety and efficiency of the aircraft.

The plane eventually became better known to the public, but much of the equipment used, and its capabilities, didn't. We had to keep quiet about it for so long that when I was finally able to tell people a little about what I used to do, they hardly believed it at all. That doesn't bother me much. What used to bother me was the attitude by some military people, that tech reps really weren't necessary after the first few years since the Air Force could just check out its own people. Well I know just how much they did need us, or else I wouldn't have been there fifteen hours a day sometimes trying to keep inexperienced people from making mistakes. I was too old to care about the politics and old enough to know what my responsibilities were. I loved that airplane like it was my very own, and in the end, I knew my corporate knowledge was valuable to the key people who had to make the command decisions.

I finally returned from Okinawa, and after a short stay at Beale, I was reassigned, and my wife and I went over to DET 4 in England. We had no idea the program would be terminated within a year.

There came that sad day when DET 1 had already been closed and DET 4 was flying its final operational sortie. I'll always remember that one with pride because I felt I was able to make a difference. The crew had just taken off when they experienced a problem with the auto-navigational system. This meant an immediate return to base. I don't imagine the balance of world peace would have been overly affected had that sortie not gone that day, but it was important to us to fly that last one. As the jet pulled into the hangar, we knew that we would only have minutes to determine and fix the problem if the sortie was to be re-launched. The tankers which were orbiting far out over the ocean, could not stay there forever.

It seemed obvious that the problem was with our autopilot inputs. After a quick analysis, I could see no faults in the system. Experience told me that the ANS could be the culprit. No one in the ANS shop agreed with me, but in the spirit of mission oriented teamwork, they checked the ANS tape. The problem was with the ANS. It was giving bad command inputs to the autopilot. We swapped out ANS units, re-launched the jet within minutes of its no-go time, and successfully flew the final SR-71 DET 4 sortie. I knew my system pretty well after all those years.

Soon thereafter, the program was cancelled and Honeywell informed me that my position was terminated since the jet was being retired. They said they really didn't have another position for me since I had been working on an "old" system which was fairly "outdated." After having been with the first completely digital system that was operational with the Air Force for over twenty years, I just had to laugh. I was 69 years old and figured I could retire anyway.

So many people knew so little about this airplane. I remember some guy bragging in the officer's club at Kadena that he had just ferried an F-15 over from the States and did it in something like fourteen hours. I told him it was about four hours in a real jet.

This plane was such a large part of my life. I know that many of us would come back to work on it again if given the opportunity. We were working on the best aircraft in the world.

— Chuck Wiethoff —
Tech Rep, Flight Control Systems
21 years

Good view of spike. Vents on nacelle are for forward and aft bypass doors.

POST STRIKE PLANS

From the Front Seat . . .

DAY TWO
1405

Walt and I are both anxious to hear how the F-111s fared. We finally get the word that in fact one F-111 was lost to enemy fire. I feel it is unlikely that the crew survived the hit and Walt agrees with me. We don't say much to each other and just deal with that news in our own quiet way. We are concerned with how many more flights we might be making.

Once we get the word that we are going back in tomorrow, we know that we will be the primary jet. There will be no strike force going in before us this time.

A new crew was flown in from Beale, and it was one of the most experienced crews in the SR-71 community. Bernie and Denny had both taught Walt and me a great deal when we had gone through our initial SR-71 training, and we liked them not only for their expertise in the airplane, but for the straightforward way they could communicate that expertise to the younger crews. Flying this airplane required a conglomerate of techniques not printed in the flight manual. Each pilot had to find those techniques which worked best for him. Passing this knowledge on to the new crews was vital, and somewhat of an art form, and Bernie was one of the best. Bernie understood the concept that there was book knowledge and then there was having a certain feel for the jet, and putting them together was what kept you alive. This was something most guys in fighter squadrons learned quickly, and when it came time to go to war, these were the kinds of guys you wanted on your wing. I was glad to see Bernie and Denny in England.

These two guys had a great sense of humor too — an important ingredient to any squadron. Two years earlier, when Walt and I were slugging it out in our simulator training, Bernie and Denny had made learning a little less painful and introduced an element of fun — something we always remembered. When I would perform some difficult action correctly in the front seat of the sim, Bernie, sitting directly behind me in the instructor's seat, would stick a wooden pointer into my cockpit with a piece of candy affixed to one end. He would gently drop the candy on the console to my right and the stick would disappear. When a less-than-brilliant action was seen emanating from the front seat, a hand would enter the cockpit and remove the candy.

Walt and I tried to play a joke on Bernie once but it backfired. Bernie had been tasked to fly a jet from Beale to March AFB for an air show and Walt and I were the mobile crew that day. Walt had this terrifically real-looking rubber snake that he thought we should put in Bernie's cockpit. Walt had already gotten me really good with that snake, placing it on the seat of the mobile car as

I was about to sit down, so of course I thought it was a great idea to get Bernie. While setting up Bernie's cockpit, I carefully placed the snake across the circuit breaker panel on the left console. This way, he wouldn't see it when he was strapping in but would find it when checking those troublesome circuit breakers behind his left arm, prior to engine start. We didn't worry about the snake making the flight, as it was a good sized piece of rubber and would easily be seen. Well, as we waited in the mobile car expecting to get some response on the radio from Bernie concerning a certain rubber reptile, nothing was said. He took off normally, and we simply thought he hadn't found it funny enough to mention. A few hours later Bernie called the squadron to let us know the jet was all bedded down for the air show. We couldn't resist asking him about the snake, and to our surprise found out that he had never seen it. I asked him how he could possibly miss it when he checked the circuit breakers. His reply was a classic: "After a thousand hours in this jet, do you think I check those circuit breakers every flight?" He had honestly never seen the snake on the whole flight. You had to love a guy like Bernie. We later did hear that the crew chief who was doing the post-flight checks after Bernie's flight did get quite a shock.

The routing for the second mission over Libya looked similar to the first day's flight, so our mission planning did not take quite as long. Apparently, our photos were receiving some very high level interest. The degree of accuracy our jet can obtain in photographing large areas at great speeds has always impressed me.

Bernie Smith. Still smiling after 1,000 hours in the jet.

We hear that the Libyan air defense was fairly quiet for the SR-71 on the first mission. After studying our routing, we note that we will be entering Libya near Benghazi and, after several turns, will be exiting the target area at Tripoli. We try to get the refueling altitudes lowered in the Med as it was something like riding a bucking horse having both burners lit while hanging on to the boom. The Greeks and Italians who control the airspace, say no, they will not vector civilian aircraft around a refueling operation, so we are stuck with the higher altitudes.

We will stick with the same jet, 960, since with minimal maintenance time there are likely to be some little things which cannot be fixed, and it is better to keep the crew with corporate knowledge in that tail number. I don't know how our maintenance people are keeping up with the tasking. We are asking them to ready both jets two days in a row — something rarely done. We are also asking for everything to work correctly on both of them. I know those guys in the hangar will be out there all night. I am glad that 960's engines seem to be humming just fine for us. The inlet door has me concerned, but as long as those J-58s are cooking, I can figure out a way to deal with the bypass door.

1900

After returning to my room, mildly fatigued, I find it is very interesting watching Margaret Thatcher on British TV. She is staunchly standing by President Reagan in her approval of the strike flights launching out of England. Many Brits are afraid of reprisals by terrorists and say there should be no more SR-71 flights at all. I admire the Prime Minister. She is holding her ground in the face of much verbal confrontation by members of Parliament. Many of her most vocal critics come across on television as spineless men, living in fear of the very terrorists Ms. Thatcher now chooses to confront.

I find it hard to get to sleep again, as the war-like atmosphere surrounding the base causes too many thoughts to continue playing in one's mind. There are protestors outside the main gate and there is some fear of terrorist reprisals aimed at our base. Outside our rooms stand military guards with M-16s. I pray our leaders are making good decisions.

From the Rear Seat...

Day Two
1410

We had flown for 3.2 hours on our first mission. The primary jet logged nearly twice that much time. Once everyone was down, we all met in the briefing room to get the latest briefing. The initial word was that three of the F-111s were unaccounted for. Within hours this changed to two aircraft missing. Finally, we got some accurate information. One of the fighters had diverted into a base in Spain with a generator problem. One F-111 had been lost to enemy fire. I knew this was relatively good in the realm of predicted losses, but that comforted me little when I kept trying to imagine who it was that went down, knowing full well that the odds of them still being alive were slim to none. I later learned that both crew members in the '111 had perished. The loss of one crew in the scope of this operation, I knew, militarily would be little more than a footnote to the event as a whole, but this hit a little close to home for me. We had been very used to flying SR-71 missions during officially peacetime operations, but the fact that we were in a shooting war had now been dramatically emphasized to me personally.

1450

We are unsure of what will happen next and wait for some word. We wondered if we would continue to fly reconnaissance missions over Libya now that the strike was over. Of course, we aren't sure if the strikes are over or not. We finally get the word that we are going back in tomorrow to photograph the target area. Apparently large dust clouds from high surface winds obscured certain parts of the area that the photo interpreters were very interested in seeing. We would be the primary jet this time, and we begin again with the mission planning.

The other crew said they experienced little in the way of hostile threats while over the target. We knew to expect a different situation now that the element of surprise was gone, and the enemy had had a whole day to prepare their defenses for a possible reattack.

The professional differences and verbal clashes that had ensued with the other crew on the first day had escalated as mission tensions rose, and the commander had wisely requested a third crew be sent over from Beale. Upon their arrival, they replaced the original crew and would fly the backup jet on the next mission. The crew that was relieved, were both extremely competent and highly experienced SR-71 crew members — knowledgeable aviators who had taught us much in our simulator training days. They simply weren't meshing as a team. There was precious little room for a personality clash at Mach 3.

The new crew arrived with a host of questions. First, though, was the problem of simply finding them a place of lodging. With all the tanker crews that had been sent over in support of the operation, there were no rooms anywhere, so these guys just slept on the couch in our rooms. These guys were two of my favorite SR-71 people — Bernie Smith, and Denny Whalen. Bernie was one of the senior pilots in the program and had been at Beale quite awhile. Denny had taught me a great deal during my simulator training at Beale and was a highly experienced RSO.

We were handed the mission maps, and they looked quite similar to what we had seen earlier that morning. We mission planned well into the afternoon.

1930

I was tired but still found it difficult to unwind and get some rest. It was a little out of the ordinary for us to fly two consecutive days. Normally after a flight, it took an evening to unwind, but now we didn't have that luxury. Brian and I both watched the British TV in our rooms with great interest, as there was great debate between Margaret Thatcher and a large segment of Parliament over the political implications of American jets launching an attack from British soil. I guess someone finally noticed the bumper crop of KC-10s on the ramp.

At this point, the big concern by the Brits was terrorist attacks in England by Libyan sympathizers. Mildenhall itself was in a condition of armed alert. We had guards posted around our building, and the base, normally pretty accessible to civilians, was shut pretty tight. The incredible state of agitation both on the TV and around the base, did little to ease my mind into sleep.

Aircraft is prepared throughout the night.

THE CRITICAL FORCE . . .

I first came to Beale in the early days of the program, back in 1965. I was the field rep for the David Clark Company of Massachusetts, the outfit that made the space suits worn by SR-71 crew members.

That company was a pretty interesting business in itself. David Clark was quite a guy — never got past the eighth grade in school, but he was an extremely intelligent self-made man. He was good friends with the Mayo brothers, who founded the Mayo Clinic, and he started out by manufacturing specialized medical equipment for them, primarily rubber tubing and bellows. The company later expanded into full manufacturing of women's girdles and brassieres. When World War II started, the assembly line was made up of long lines of women stitching those garments.

During that time, a requirement came down from the Department of Defense which would change the face of the Clark Company forever. Pilots of the newer and faster fighter planes were experiencing increasing cases of blacking out when pulling out of steep dives. This was, of course, due to the gravitational forces pulling the blood down toward the legs of the pilots. The David Clark company was tasked with devising some sort of inflatable anti-gravitation garment, that would, with the onset of high G force, constrict the pilot's torso and legs so as to slow the downward flow of blood. This would help to keep him fully conscious throughout high G maneuvers. As you know, the G-suit is now a required item in the fighter pilot's arsenal of equipment. David Clark saw great potential in this type of product for his company and began hiring ex-Army Air Corps people with some knowledge of what was really needed in the field. When I was hired, not too long after the war, Mr. Clark told me that I could expect the unexpected if I worked for him, and he wasn't kidding.

The company eventually got into the pressure suit business and was the prime supplier to the Air Force for the suits used in the U-2 and the X-15 program. David Clark would take me to lunch sometimes and discuss all kinds of manufacturing ideas that he had for the future. One day he asked me to go down to the airport to pick up a military flier who was coming in to be fitted for a suit. When I got to the airport, I found out that it was Gus Grissom I was meeting. He was already famous at that time and turned out to be the nicest guy, very easy to talk to.

During the same period, folks in the company knew that there was some other program getting a lot of interest, but no one knew what it was. It seemed there were a lot of Air Force types coming and going, and we usually picked up on these sorts of things. We were sure it had something to do with the Air Force Dyna Soar program but it didn't.

David Clark called me into his office, closed the door, and explained to me that the Air Force was adding a new plane to its inventory, designed to fly faster and higher than any other plane,

Diagnostic checks are run prior to every flight.

Maintenance crew checks jet on ramp where it will start tomorrow.

The space suit: a critical garment.

even than the U-2, and its crews would need space suit protection. The challenge here was that the suits being used on the X-15 and U-2 would have to be modified extensively to meet the new aircraft's specific requirements. Mr. Clark asked me if I would like to be the project manager for the integration of the space suit to this new plane. I said yes, and I hadn't even seen a picture of what the plane looked like. Of course it was the SR-71 and I was anxious to see it in person.

One of the first things I got to do was go out to the ranch and finally get a look at this new plane. The program was very well thought out. I flew into Los Angeles, made a generic phone call, and a short time later some people would pick me up and fly me out to Groom Lake. I knew I was definitely a part of something very important. Upon seeing the plane for the first time, I instantly knew that I would be working with something very different. It was awe inspiring. I had seen a lot in my days in the military, but nothing like this. The plane itself had such a presence.

Sometimes, we would make suit modifications right on the spot. Like everyone else, we didn't want our area of responsibility to be the cause of an abort, and we would work many late hours to get everything just right for the next day's flight. The Air Force had outlined some basic requirements for the suit on this plane. In addition to maximum safety for the crew member, the suit was to be as comfortable as possible for long missions in a small cockpit. We developed twelve different patterns for suit sizes so that we could custom-fit them to each crew member. The new patterns were much better than using common sizes like small, medium or large, as had been done with the U-2. This took about sixty separate measurements by our people back in Massachusetts, but it really made a difference to the crews.

We also found out during that first year that some sort of inserts needed to be worn inside the gloves. Some pilots' hands were just getting too sweaty inside that rubber glove. So thereafter, crews could wear thin white gloves inside the suit, which gave them better grip and actually did feel much better on their skin. Someone came up with the idea of writing critical bailout procedures in bold letters across the sleeve of the suit. It was one of those ideas that just stuck for the life of the program. Also, since the airplane was equipped with an ejection seat, the suit had to be stitched with parachute harness connectors and a lap belt connector. The parachute was housed inside the seat, so the crew didn't have to carry it to the plane, but they had to be firmly attached to that seat and chute for it all to work in case of ejection. Stitching the connectors onto the suit had to meticulously be accomplished without error. Much like sewing a parachute, if those harnesses on the suit had one wrong stitch, off it came and the seamstress had to start all over again.

Additionally, we had to figure out a way to include some sort of water flotation device for the crew. It was a pretty fair assumption to think that this plane would be flying over great stretches of water during its missions. If a crew member bailed out and hit the water without a life preserver, it would be pretty tough for him to stay above water with all that weight. It took us several tries to get those LPUs to hang on the suit comfortably. It seemed like we were hanging so much stuff on the suit that the crew wouldn't be able to move, but somehow they managed. We were very concerned with keeping as much mobility in the suit as possible.

We also got involved with the development of the urinary collection device. We looked at many alternatives, and most were simply too bulky. Initially the UCD was to be inserted inside the suit, but this was too uncomfortable and we settled on a workable external device. It consisted simply of a catheter-like device which attached to an opening in the undergarment. On the tip was an opening which attached to tubing which ran down the leg of the suit into a small collection bottle kept in the leg pocket. If the crew member had to do anything else, he was in trouble.

The helmet was really an excellent design, and we had very little trouble with it over the years. It was designed to give the crew member one and a half psi greater air pressure than the rest of the suit, so that he wouldn't have to breath his own exhaled air.

The face heat on the inner visor proved to be an interesting development also. During the early test days, Lockheed had lost a jet when it came apart at high Mach. The backseater was killed, but the pilot found himself falling through space with a helmet visor which was completely frosted over. He knew if he opened it, it would kill him at those altitudes, but he needed to see what was happening. Eventually he opened it up and right about then his chute deployed, which meant he was at a safe altitude. He was lucky, and we knew that we needed to devise an emergency face heat system for bailout. Again, finding the room for more equipment was the biggest problem. We eventually were able to squeeze a small battery into the survival kit built into the seat. Of course, your rubber raft was in there too, so we had to be careful.

Oddly enough, the David Clark Company ended up manufacturing the life rafts themselves. In fact, as time went on, and other contractors were unable to continue to meet our requests for small quantities, the David Clark Company increasingly made everything for themselves. Most people know that we provided the space suit but don't realize that we eventually also manufactured and supplied the helmets, the glass visor for the helmet, the life rafts, and the entire glove assembly.

When I came to Beale in 1965, one of my primary jobs was to set up a training program for suit technicians. Originally, all the suit specialists had been civilians, but now the Air Force wanted to train its own people. Most of the people I saw in those classes were top notch individuals but had no experience with a suit of this type. These folks learned the system well, and it is a real credit to them that for the life of the program the Air Force was able to take over the entire operation. And we never, ever, lost a life due to a suit malfunction. It was an extremely reliable piece of equipment.

One time, I remember Pat Halloran was sitting in the plane ready for takeoff and for no apparent reason his helmet visor cracked. It cost us the mission, as he wisely taxied back to the shelter, but that was a freak occurrence.

Walter suits up.

102 THE UNTOUCHABLES

The color of the suits changed over the years, too. Originally the silver suits were designed for the X-15 with the idea of keeping the pilot cooler. Then we went to white. The problem was that the glare from the silver and white suits was distracting to the pilot trying to read the gauges. Looking into the dials on the dashboard, pilots would frequently complain of seeing nothing except the white reflection of their suit. We originally attempted to dye the suits a darker color. This absolutely didn't work. The Air Force wanted a suit made of the flame retardant material, nomex, anyway, so we fabricated the newer "gold" suit in the '70s, and that one has lasted.

It was enormous job satisfaction to be part of such an interesting program. That suit was such an integral part of the mission that I really felt close to the crews. Up until I left the program in the late 70s, I knew every crew member as they had to come to me for their suit fitting and training. These guys were terrific to work with. They all had different personalities and some were real comedians. I saw many of those early crew members go on to be generals, too — even some of the ones with a sense of humor.

I guess for me, what made my whole association with the program so special was the people. Both the Air Force and civilians alike were very professional and quite dedicated to the mission. I helped suit up some pretty well known people too. I remember when Barry Goldwater got a flight in the B model. He was very pleasant and extremely knowledgeable about the plane and the suit.

Near the end of the program, DET 6 called the David Clark Company one day and said that the Smithsonian Air and Space Museum wanted one of the SR-71 space suits for permanent display. I personally picked one out and shipped it to them. I recently was in the museum and saw that suit on display. I was very proud seeing it there. I felt so much more than I could put into words, seeing that suit and knowing all that had gone into its creation.

Having been a part of that program was truly one of the highlights of not just my professional career but of my entire life. When the jet was terminated, I honestly felt like I had lost a good friend.

— Bob Antilla —
Tech Rep, Space Suit
16 years

" If you've seen one Mach, you've seen them all."
    Lt. Col. Buddy Brown, in response to the question, "What is flight at Mach 3 like?"

CHAPTER 4

The Mission

From the Front Seat...

16 April, 1986
Day Three
0415

We sat through the morning briefing quietly, most of our questions having been answered yesterday. The commander briefs us on a non-related SR-71 mission out of Beale. One of our buddies coming off a Cuba run apparently developed engine trouble and had to make an emergency landing at Key West. The engine was burned beyond useful repair, but it got them down OK. Walt hates hearing stuff like this just before going to fly. I tell him, maybe it's better in a way to hear this news now, since the odds of it happening again so soon are pretty low, so he shouldn't worry. This is too unscientific for Walter and he gives me one of those engineering looks. There are so few countries supporting us in this operation that I genuinely hope we do not have to make any unscheduled landings along the route.

The threat of Libyan MiGs or SAMs launching against us is very real, and Walt and I take a few moments to discuss some indications he may see on the DEF panel, and some actions I may have to take. We feel slightly less concerned with the air threat since, in the past year, we had received good intercept data from our own Navy and Air Force.

On training sorties out of Beale, we had frequently flown across the air combat ranges in Nevada. Once over the range, F-15 or F-14 aircraft attempted to intercept us. These missions were always precoordinated and yielded good information about the inherent difficulty of trying to bring down a target travelling 2,000 miles per hour at high altitude. Even though we provided the interceptors with our exact course, speed, and altitude, they still had difficulty acquiring a valid lock-on. I regarded the F-15 as the world's finest air-to-air platform. In talking with some of the Eagle pilots, I learned that they could get close, but it still would take a fair amount of luck to actually hit us, even when we were giving them our parameters. One time, we had to cancel our scheduled flight over the range with an F-14 squadron. When we called back to their squadron the next day to reschedule, their squadron executive officer thought we were calling for the results of the mission and promptly informed us that the Tomcats had scored three sure "kills" on the SR-71. I was impressed with such squadron loyalty but had to inform him that we hadn't even taken off.

Ever-present JP-7 leakage.

TEB light on start.

THE MISSION *107*

The Nevada ranges had also provided us with valuable data concerning surface-to-air missiles and their ability to track and lock on to our aircraft. We were more concerned with this threat since the data had shown us that we were not as invulnerable as everyone liked to think.

0530

Prior to getting in the jet, we have some brief conversations with some of the tech reps in the hangar. They just want to let us know about some of the little things that they couldn't really spend time working on, so we are to expect some minor problems in flight. This is understandable considering the work load they have been under.

I reassure Doc, the engine man, that I'll be kind to those J-58s, since they are one of the few components of the aircraft giving us no problems at all. Doc proceeds to tell me about some work they have done with the engines last night, but even more reassuring to me than his words is the caring look on his face. I know he treats those engines like a father would his child. I also know that he is genuinely concerned for our safety. It is a wonderful thing to be able to talk to people like this as the last thing we do before entering the cockpit.

I give the jet an extra pat on the nose to let her know I forgive her for all the vibrations yesterday, and to remind her that she is the primary bird today and it's time to show her stuff.

Max 'burner on takeoff roll

Once inside the cockpit, I notice directly before me a small picture taped to the map display screen. It has been cut out of some British tabloid and shows a well-endowed, scantily clothed young lady . The freshly penciled-in words, now emanating from her mouth, suggest she knows me intimately and that she has somehow been briefed on today's mission. It is a tradition, and no matter how many times we saw these remnants of creative mobile crews, they always made us laugh.

The engine start goes smoothly and the crew chief hands us a message from the mobile car relaying that the barrier at the end of the runway is inoperative today. This matters little to me and will have no affect on our mission. With the heavy fuel load, I know that trying to abort 60 tons of titanium and steel on a wet runway is risky at best. In the event of a catastrophic high speed abort, I placed more faith in the yellow rubber handle between my legs marked EJECT. Subconsciously, I address these thoughts on each and every flight, and then push them aside to concentrate on keeping up with the takeoff when all goes normal.

Our takeoff goes much smoother today as I remember the extra fuel load and delay my rotation slightly. Again, after slugging through the gray mist we are greeted with bright sunshine once we break out on top of the clouds. The same weather system which is producing the clear

The right console in rear seat showing ANS panel.

skies above 10,000 feet is also producing some moderate turbulence. This is quite undesirable in the refueling track.

I'm hoping we get a tanker with a good autopilot this time. We don't. It turns out to be the same exact tanker, with the same exact problems. The tanker maintenance crews are fighting the same battle as our maintenance folks; no time to fix nonessential items, and no spare planes to put up. We're all in the same planes, and we experience the same roller coaster ride on the boom, now with the addition of some healthy turbulence. With one menacing gust of angry air, I see the boom flex wildly and disconnect from our plane. Like a severe bump in a road, the turbulence has bounced our jet precariously close to the belly of the tanker. I push hard forward on the stick and bank the aircraft sharply to the right to avoid a collision. The slight nose-down attitude of our jet has given us enough extra knots of speed to move us dangerously forward, beneath the tanker fuselage. I then find myself straining to turn my space helmet and peer over my shoulder to keep the tanker in view. I am peering into a fireball of a sun low on the horizon, as I move my very heavy aircraft aft and, very carefully upward to regain the contact position. At 300 knots, we are within ten knots of our minimum airspeed for this weight. Trying to "back up" with a reduction in power, is not comfortable, and I continue to hope there are no more "bumps" in the road. The whole process feels like a good workout at the base gym, except less fun.

The extra maneuvering has cost us some time, and Walter has coordinated and gotten us a track extension so we can have the time to fill our tanks. Walt's calm voice in the back is a welcome change from the sound of my own heavy breathing.

With my faceplate starting to fog up, I finally, literally, drop off the tanker, and gladly begin the acceleration to the high blue. The familiar nudge in my seat as the 'burners light off is a welcome feeling. In a few moments the tanker is well behind us as we push past 400 knots and continue to accelerate.

Walt reminds me that he is going to be busy running some extra tests on the cameras so he won't be backing me up as much with the numbers in the climb. This is not much of a problem as the adrenaline is high for this mission and attention to detail is intense.

As the speed moves through Mach 1.4, the inlet door begins vibrating on cue. It is bothersome and noisy but, in itself, should not stop the mission. At 1.6 Mach, the spikes begin their slow journey aft into the nacelles (one and five-eighths inch per every tenth of Mach number). I monitor them carefully to ensure they are maintaining equal positions. A divergence of spike position is the best way to incur an unstart and that would cost us precious fuel, something we can ill afford this day.

Passing through Mach 2.5, the door vibration increases slightly and I don't like what this may mean to us in the way of drag. I run the jet on out past Mach 3.1, and although the bad inlet door sounds like things are coming apart, everything is holding steady.

Thirsty Sled approaching tanker.

Tanker navigator plots rendezvous with SR-71.

Boomer's view of contact.

112 THE UNTOUCHABLES

I cross-check my triple display indicator with Walt's. The TDI is my primary reference for equivalent airspeed, Mach, and altitude in the regime we'll be flying. The normal barometric pressure instruments in the cockpit, though still registering, are simply too inaccurate at these speeds. The TDI compensates for compressibility errors and produces amazingly accurate readings for the pilot. Walter's TDI agrees with mine, as it should. By the time I have leveled the aircraft off at 73,000 feet, we have burned nearly 20,000 pounds of the fuel our first tanker gave us. Once I see the target Mach, I can pull the throttles back a bit and start to get some pretty good mileage out of a plane that cruises in afterburner with ease.

Walter alerts me to the fact there is a problem with the main camera. There is nothing I can do about this, except to ask Walt if it's a go or no-go for the mission. There is a page in our checklist which spells out specifically what systems are needed for a mission to go. It is simply a go, or a no-go; there is no free-lancing by the crew in these matters. Walt tells me for now it is still a go. I don't know of anyone in the squadron I would rather have in my back seat right now, trying to troubleshoot that sort of problem. I feel confident that Walter will come up with something, as he always has in the past. I can't worry too much about back seat problems now, as I am more interested in the warm outside air temps which are pushing us below our programmed fuel.

Speeding southward across the intense blue of the upper stratosphere, I settle into my familiar routine of cockpit management. Fuel tanks are checked to ensure the proper pumps are feeding.

A quick glance outside at Mach 3.1.

Mach is adjusted with small movements of my left hand on the throttles. My right hand rests on the right console with one finger caressing the autopilot pitch wheel as I will constantly be adjusting our altitude throughout the flight. I scan temperature gauges. I cross-check the center of gravity reading with what Walter's CG says. I know from past experience that if my gauge has failed, the needle simply freezes at the last valid reading and I might not know my gauge is in error until the aircraft starts making drastic auto-pitch changes to compensate for being off balance. I check the pitch indicators and they confirm we are running clean with minimal drag. I reach above my helmet and push the small periscope up through the cockpit hatch. It gives me a rearward view, and I check the rudders to ensure they are trimmed flush with the aircraft, again to reduce any drag. I glance at the SAS panel and see no lights. This is good. I try not to look at it too much as this sometimes causes warning lights to appear. Double sets of engine gauges are checked and they are rock solid. Not so much as a quiver of one psi on the oil pressure gauge.

I love the J-58, and the reassuring consistency of its performance is contrasted now by the continued fluctuations and vibrations of the inlet doors. The marriage of the inlet system and the J-58 was vital for this jet to achieve the performance it has. It is a work of genius and no system was more difficult to develop in the building of the SR-71; and no system gets more attention from the pilot on every flight.

High above the Atlantic.

Front cockpit.

Pilot self-portrait.

The quiet intensity of my cockpit is interrupted by the familiar voice of Walter telling me the descent point for our next refueling is upon us. It seemed to come up so quickly. I slide the throttles back over the detent, and the jet shudders slightly as the afterburners are extinguished and we start down.

Again, the Mediterranean looks exquisite and I am relieved to see the clear weather. I know we are coming close to numerous civilian airways as we descend and I keep my eyes out of the cockpit more than normal. Walter says he is ranging on the tankers, so we know they are out there.

On the way down, the plane growls like a wounded tiger. She always hates coming down. I have not helped her pain today as I have neglected to manually select additional fuel pumps in the descent. This helps mix some of the cooler fuel with the hot fuel. It is just a technique but one I normally remember when not so preoccupied with rough inlets and high temps. She growls at me all the way down as hot fuel sloshes throughout three of her six tanks.

Once below 50,000 feet, I always tended to keep my eyes outside more than inside in order to check for any other air traffic. The SR-71 was not equipped with any type of weather or search radar, so normally we were reliant on air traffic controllers to paint us on *their* radars and call out conflicting traffic to us. Now, in a comm-out environment, we had no "eyes" except those of the guy in the front.

Front seat — left console.

Passing through 38,000 feet at about 400 knots, something on my left window catches my attention. I think it must be a bug and as I turn to look at it, I see, very clearly, the brightly colored fuselage of an Alitalia 727, so close to me that the windows appear the same size as mine. The airliner passes dangerously close to our nose, and in the time it takes me to move my head, the near-miss is over. That was the closest Walt and I ever came to hitting anything (the occasional tanker excluded) and to this day I have every close-up detail of that aircraft firmly etched in my memory. My witnessing of our near collision, and the speed at which it happened, stunned me into silence. By the time my pulse slowed to subsonic, I didn't think it would make Walt feel any better to tell him about it.

From the Rear Seat . . .

Day Three
0550

Finally it is time to settle into the familiar confines of what will be my office for the next six hours. Though the mission is intensely different from others we have flown, I do almost nothing different from my normal habit patterns. I guess that was one advantage of flying regularly toward hostile environments; we now had little to adjust for a mission of increased intensity, and that really helped.

The only anticipation I did feel was thinking that Qaddafi certainly would have no hesitation in firing on us today. This is where the performance of the jet was so important. If a bad inlet or an engine problem caused us to get slow or low, we could become quite vulnerable. With these thoughts, I ran through the DEF self-test, and it tried to impress me with its usual light show telling me that it was working perfectly — but I always had some reservations about relying too heavily on this type of equipment.

The main sensor for the mission was the optical bar camera. This meant labor-intensive attention by me since all the camera's settings were done manually from my cockpit. The camera was housed in the nose section and would provide horizon to horizon coverage. As a backup system, the aircraft was also loaded with the smaller technical objective cameras, housed in bays just forward of the wing. Aside from my simply checking their settings, these cameras were set to run automatically with inputs from the ANS tapes.

As the PSD crew was handing me the last of what seemed like a small library of checklists, comm kit, maps, and mission summaries, I was already checking the ANS to ensure that all the times and data were correct. They were.

I then checked the backup inertial nav system. This small INS was typical of the types of nav systems that most fighter aircraft used as a primary navigational tool. For us, the INS was a nice

RSO's office.

backup system to the powerful astro-inertial nav system. I used the smaller INS to program in the coordinates of various divert bases that we might use along the route. This way, in case of an emergency I could quickly call up a location and give Brian a quick vector to the most suitable point of intended landing.

Right in front of me sat the projected map display. In any given mission, this was either my best friend or a major pain in the butt. The system provided me with a moving map display of the route based on programmed speed and time of flight. If there were errors in the nav system, or if we deviated from the course line, the map just continued projecting where it thought we should be, not where we actually were. Normally, as long as it stayed synchronized with our speed it worked fine, but the other headache with it was that there was usually so much extra printed data on the maps from the planners and the intell folks that it was a little like reading the fast credits at the end of a movie with a cast of thousands. Info printed alongside the route would normally include air refueling data, descent points, divert bases, true and magnetic headings for each leg of the route, and some radio frequencies, just to name a few. Sometimes it was a science just trying to find that little black line we were supposed to be flying along. Since this was a completely new route and the film cassettes were literally fresh off the press, I was happy to find the map was actually less cluttered than usual and quite easy to read.

Just below the map display was the optical viewsight. This enabled the guy in the back to see directly below the aircraft via a viewing lens mounted flush on the bottom of the jet. This provided me with about a 200-square-mile footprint across our route. The guy in the front always thought he had the best view of everything, but he could only see out where we were going; I could view exactly what we were flying over. It was a little like looking through a bomb sight, albeit a higher view than most bombardiers would ever see. The viewsight had scales and a sighting device and could be used to update our precise position or get a view of where the cameras were looking. The temptation to watch this picture for more than a few minutes in flight was quickly squelched, as doing so rapidly developed into a motion sickness maneuver.

Though not really designed for this purpose, the viewsight had come in handy on more than one occasion when we had to make approaches for landing in particularly bad weather. With the jet in the landing configuration, the pilot sat in a fairly nose-high attitude and occasionally had a difficult time picking up the approach lights during low visibility approaches. By turning the viewsight on, the RSO could see directly below the jet and would normally see the approach lights first, a good sign that the jet was lined up with the runway even though the pilot might only see solid gray weather. Being only a couple hundred feet above the ground at slow speed with nothing in view, it was a "warm fuzzy" for the pilot to hear that the RSO had approach lights directly below the jet.

Two good 'burners.

Gear up.

Everything was checking out fine in my cockpit. I reviewed the timing coordination necessary with the second plane so I would be sure to make the appropriate radio transmissions when necessary.

Finally I placed my two little tube food containers off to the side. The little shot of food in flight wasn't so much because of intense hunger; I used it more as a calming device to slow me down when I was getting too busy or worrying about something I couldn't control. Usually I ate something while Brian was grunting and groaning on the end of the boom during the refueling. It helped me not think about just how close we were to that tanker. With butterscotch pudding, some apple sauce and my Gatorade in place, my cockpit was secure and I was ready for engine start.

Barry MacKean, the DET commander got out of the mobile car and gave us the signal to start. After going through the normal start procedures, Barry did something a bit out of the ordinary. Prior to the canopies being closed, Barry came up the ladder, reached in, and shook hands with Brian and myself. In my entire time in the program, I don't remember that ever happening before. It was a nice touch and we appreciated Barry's gesture more than he ever probably realized.

We were the first SR to launch that day, and Bernie and Denny would follow us about an hour later. We cruised out the same route we had flown the previous day without any radio transmis-

KC-135Q with boom extended.

THE MISSION *121*

Walter's checklist.

The backup jet is readied.

122 THE UNTOUCHABLES

sions. I hit the appropriate radar identification codes to electronically let British radar know our position so they could deconflict us with other air traffic.

The refueling track is rough and I know Brian has his hands full. As I am mildly tossed about in my seat, I check the ANS and see that it now has picked up some stars and is working perfectly. I wanted to be sure since, when I checked it on the runway, the weather at Mildenhall had precluded it from tracking anything. An abrupt change of flight attitude results in a slight negative G feeling in my seat and our disconnect from the tanker. With no forward view, I can only imagine what Brian is seeing. It is a rough day. I have faith Brian will get the gas. I suck down half of my Gatorade.

Normally, the times of intense work load for the RSO and pilot were on opposite ends of the flight regime. Subsonically, the pilot was working the hardest with takeoff and landing, refuelings, and basically hand-flying the jet. Once leveled off at high speed, with the autopilot and auto-nav engaged, assuming a good jet, his work load was eased to smoothly controlling Mach and altitude. For the RSO, subsonic time was the opportunity to run some checks and relax a little. Once up at speed, though, the RSO was extremely busy with the navigation, radios, sensor action points, descent points for refueling, DEF activations, position updates, and assisting the pilot as necessary.

The one time that both crew members were equally busy was during the acceleration climb to high Mach. During the accel, airspeeds, Mach, and altitudes were all cross-checked at various points to ensure that all instrumentation was accurate. Also watched closely after Mach 1.6 were the aft movements of the spikes. The RSO did not have indicators for all of these things in his cockpit but referenced the desired numbers in the checklist to the pilot throughout the climb.

Of course, as all of this was going on, there were the normal navigational duties by the RSO as certain checkpoints were passed. Typically, from the time we accelerated past the tanker, full of fuel, went through the accel, and leveled off around 70,000 feet at Mach 3, it would take about fifteen minutes, so things were happening pretty fast.

On this particular mission, I had been given a command order to recheck the OBC during the climb. I know the commanders were thinking that since this was the primary sensor, it would be a good idea to recheck it, but I felt uneasy about this idea. I tend to think of equipment in a very mechanical way, and having seen the OBC pass its checks on the ground, I had the distinct feeling that only something bad could happen by checking it again.

This was a busy time, but going through about 60,000 feet, I told Brian my head would be down and I'd be working with the cameras. Sure enough, the camera went through about three frames of film and jammed. Everyone who ever flew an SR-71 would agree that the airplane seemed to have a personality and mind of its own. The camera seemed to be telling me that it didn't like being tested where it normally wasn't. I now was faced with doing everything I could to get it to work. This required a complete shutdown and re-initiation of the system. I knew it was going to be a long day. The whole process took about fifteen minutes, and by then Brian had leveled off and the jet seemed to be flying all right, although the outside air temps weren't quite what we wanted.

I initially thought I'd have plenty of time between Land's End and the Strait of Gibraltar to get everything in order and rechecked prior to entering the Mediterranean, but suddenly the mission was racing by. I cross-check the ANS data against the INS and the TACAN. We seem to be right on the money with course. This gives me a feeling of confidence and I again turn my attention to the camera.

When the OBC had failed initially, I felt a little like I was back in the sim. I had been bombarded by the sim instructors with this sort of failure in those days, but it had prepared me well for the situation at hand. I had learned that by going through the entire setup sequence, the camera could unjam itself — maybe. I always tried to keep Brian apprised of things going on in the back seat, and I explained very briefly and simply (this was the best way to communicate with someone controlling a speeding bullet) what was happening, even though I knew there wasn't anything he could do about it. He understood the importance of the primary sensor and said he had great faith in me to fix whatever needed fixing. He was also quite concerned with the buzzing of the forward doors and was increasing the Mach beyond 3.0 to determine how the inlet would do at the higher speeds we were programmed for later in the mission. We were both busy and there

68,000 feet.

was little talking. The next time I looked up, we were abeam Portugal and starting our turn into the Med.

Slicing through the Strait of Gibraltar, I double-check our descent point for the tankers and remind myself that we aren't descending as low today, as we'll be refueling closer to 30,000 feet. I note the weather is beautiful and I take this opportunity to use some obvious landmarks in the blue sea below us, to once again verify our navigational data. Once in the sensitive area, I will turn off all non-essential equipment that emits any kind of a signal. This will include the TACAN, something likely to be used in case of an unscheduled landing at a divert base, so I give it one last check against known points before shutting it down. Looking through the viewsight, I compare the TACAN data with what the ANS is telling me, and it shows to be within a half mile — pretty good for a TACAN system that was never designed to fly at these speeds and altitudes.

Descending, reluctantly.

Approximately 200 miles from the rendezvous point, I tell Brian to come out of 'burner and start the descent toward our tankers. As the jet went through its usual sounds of a wounded banshee in the descent, I set up the radios and the proper identification codes for ranging on the KC-135Q. Winds at lower altitudes could play havoc with the rendezvous sometimes, but my ranging needle pointed just slightly left of our nose, so the tankers looked to be in the right spot. I was glad to see such clear weather, as no less than six tanker aircraft were on station to support the two SR-71s.

Brian got a good visual on the tanker early, so that made things easier as we eased up toward a KC-10 and a '135. Everything was still comm out, so initially discerning which set of tankers was ours was a little confusing. A smart boom operator lowered his refueling boom and sprayed a short burst of fuel into the sky, and Brian positioned us behind that KC-10.

Turbulent contact.

THE CRITICAL FORCE . . .

When I interviewed for a job as a civilian employee with the Air Force Logistics Command, I didn't have any idea exactly what the job entailed. A small group of us were taken down the hall in this very secretive building and asked some questions. None of us knew hardly anything about the SR-71 and its mission at that point.

I was originally assigned as an engine specialist. Now, the first time I saw that J-58, well, it was awesome. I honestly thought something was wrong with it; there was entirely too much plumbing and wiring externally. Most engines were "cleaner" on the outside. This was a very "busy" engine. They took me up to Beale Air Force Base and I witnessed an actual launch of the jet and got to see the engine run on the test stand. I was very impressed and knew it would be a challenging assignment. Each time I learned something new about the program, I was more impressed with this jet. This feeling continued for more than ten years with the program.

Having a maintenance background really helped now that I was in the supply and parts end of things. This gave me not only a good rapport with the guys in the field but a better understanding of the systems.

When I would go out into the field, which for us meant going up to Beale, I was always impressed with the pride and enthusiastic attitude of the people there. I also learned that they expected guys like me to have knowledge of many other systems besides just mine.

The jet was having more than its share of wing fires in the early days. It was obvious that the original fuel pumps needed to be redesigned. There was simply too much vibration in the accessories section. This sometimes caused a fuel leak. Now keep in mind that at Mach 3, fuel doesn't simply leak, it vaporizes. This vaporized fuel would then meet the incredible heat that was

When she was broke it was usually no easy fix.

Crew chats with technicians prior to entering jet.

Jerry Keever oversees warehouse full of J-58s at DET 6.

Servicing TEB on engine.

Static engine run at Kadena.

generated around the engine at high speed. The fuel was originally designed to operated at no more than 350°F. Upon inspection, we were finding the temperatures frequently reaching in excess of 400°F. Then some company came up with a "smart" valve that was supposed to fix that situation. Of course it never did, and we kept right on flying, which I guess was a tribute to the structural integrity of the engine as a whole. With the new fuel pumps though, we did significantly reduce the number of wing fires throughout the fleet.

We were really careful with those engines and monitored them closely. Remember that no one else was asking a powerplant to accelerate to Mach 3 and then sustain that speed for thirty or forty minutes, all while cruising in afterburner. The engine itself very rarely caused the jet to abort a mission. It usually got the blame, but more often than not it was an accessory on the engine or an inlet problem. We meticulously measured how many minutes each engine had at high Mach in addition to the normal engine times. Getting spare parts in the early days was a problem, but near the end we were in very good shape with parts.

When they were sent for overhaul, we initially couldn't understand why it took so long. We found out later that the Pratt and Whitney folks were being overly cautious with the exotic metals used in the engine, primarily the gold, and had instituted a cumbersome degree of security to

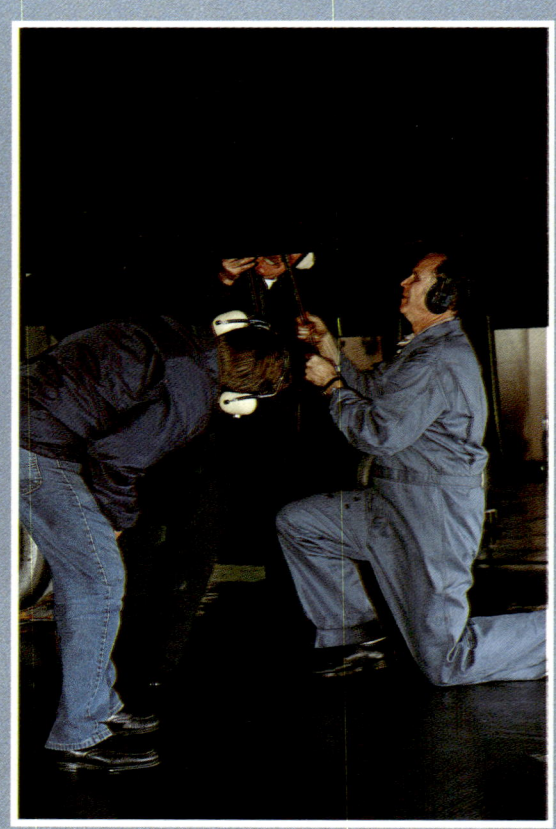

Men at work.

inhibit theft. As it turned out, many of the parts of the engine originally incorporating gold were later replaced with newer composite metals, so it wasn't really a problem.

Keep in mind that in the early days, the maintenance guys out on the flight line didn't even have the standard servicing publications that were available on other Air Force systems. What they lacked most was the illustrated parts breakdown, or IPB manual. For the maintainer, this would be the equivalent of tearing into your car's engine without the benefit of using a parts diagram illustrated in detail in the auto manual.

The root of the problem stemmed from the original thought that this aircraft would be maintained solely by Lockheed civilians. These civilians would have the kind of expertise on the system that wouldn't require IPBs. They would simply use prints from original photographs of engineering diagrams. This wasn't ever considered a problem, since the jet was designed with the idea that it would fly for about ten years and there would be no Air Force maintenance. This worked nicely when the CIA owned the first jets, but then later when the Air Force got the reconnaissance version, they very much wanted to be "hands-on" with the plane.

I can remember Air Force technicians having to laboriously refer to engineering prints, which of course were kept in a secure location other than the hangar. Eventually this was all worked out through the dedicated efforts of individuals who strove to better the system. But early on, I really was impressed with how some of these young guys took to that plane, even without sufficient manuals.

The Air Force wanted to be able to do everything, but sometimes this meant some growing pains. When it came to the TEB, they really had to ask themselves if maybe everyone on the line should really handle that stuff. The more they found out about TEB, the more dangerous they realized it was. TEB doesn't just dissipate, it has to burn off at about 2,000°F.

Well, when it came time to send an engine to the plant for overhaul, the TEB had to be purged. More than once, an engine mechanic at the plant was greeted with a small fire upon removing certain engine panels, as the TEB ignited upon contact with oxygen. Obviously, the TEB purging procedures needed special attention.

I always was surprised we didn't have more problems with TEB, considering the volatile nature of the stuff. Just about the time we had all gotten pretty comfortable with TEB in the aircraft, I received a rather unusual phone call at DET 6 one day. The caller asked if this really was the place where we handled SR-71 parts. This really surprised me since very few people outside of the program knew exactly what we did there — or so we thought. Before I could ask some questions, he proceeded, in a very serious and concerned manner, to tell me that he had published a paper on Tri-Ethyl Borane while in college and that now, as a chemist, he was very surprised to read in a professional journal that we were using it in the aircraft to ignite the fuel. Before, I could even stop him to ask just how he came upon this number, he continued on by saying that TEB had originally been developed for oil drilling companies for the specific purpose of blowing large holes in the earth but certainly not for the starting of jet engines. He said he thought we just might like to know that, and he was glad to help. Then he simply hung up. Utterly amazing. As I recall, we never shared this information with any of the crew members.

— Barry DeVries —
Engine Technician, Configuration Control
9 years

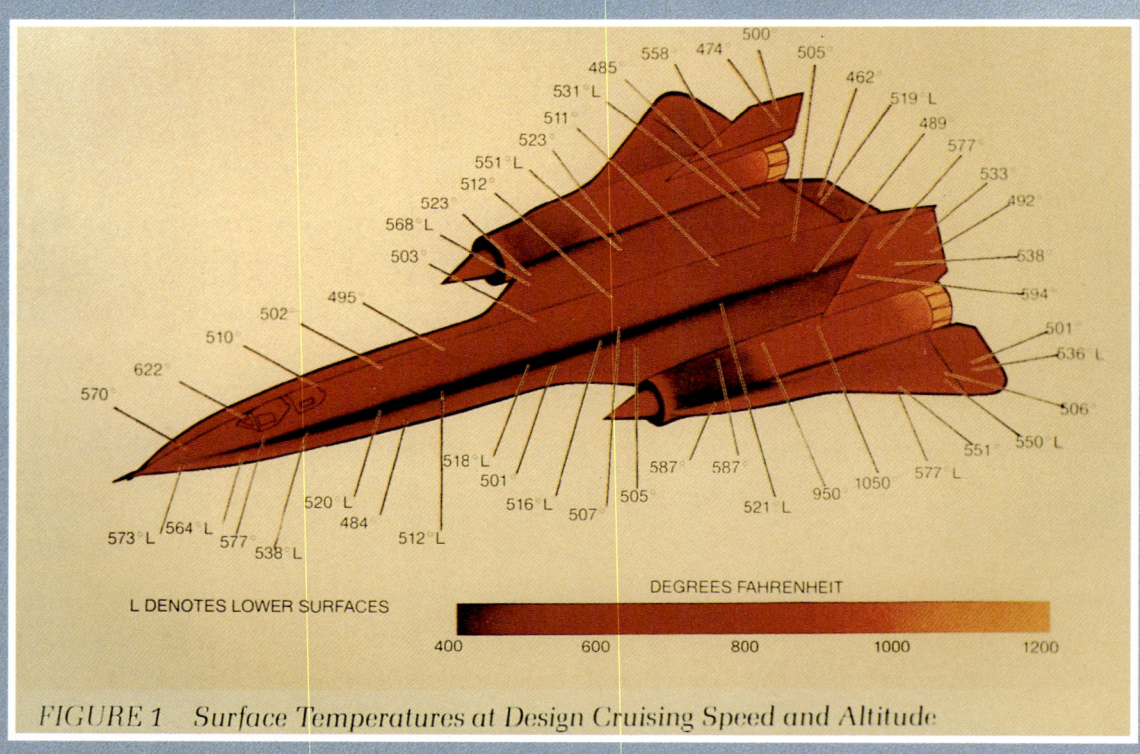

FIGURE 1 Surface Temperatures at Design Cruising Speed and Altitude

MESSAGE IN THE MED

From the Front Seat . . .

Day Three
0815

Refueling from the KC-10 was quite different than from the smaller '135. The '10 is one big airplane, and it was always a little intimidating coming up underneath that much mass. The higher refueling altitudes weren't too bad initially, since the weather was good and the autopilot in this tanker worked fine.

As I eased my jet forward, I felt the slight air disturbance across my tail caused by the small wings on the tanker's boom. To the receiver pilot, a good boom operator was worth his weight in gold, and this boomer was a real pro as he raised the nozzle slightly to allow me to slide into position without the usual buffeting. I felt the comforting nudge of the refueling nozzle entering our jet's receptacle. Just as we started to take on gas, we got a disconnect. I wasn't sure why we had come off the boom, but I repositioned and hooked up again. Again, after re-establishing contact with the boom, we became disconnected. Walt asked me if I was having a problem and I told him I didn't think so. I tried it again and this time, just prior to getting disconnected, we heard radio transmissions from the tanker.

Walter very astutely figured out that their transmissions were electrically causing the disconnects. Each time they keyed their mic button, the electrical boom latching device received a disconnect signal. Once again, I moved onto the boom, and this time Walt quickly informed the tanker crew to remain silent until we got the gas. As our thirsty jet swallowed fuel at 6,000 pounds per minute, Walter expertly adjusted enough secure radio switches to enable us to converse with the tanker without getting clicked off the boom.

This was good, because by this time our jet was bucking along at these high refueling altitudes with two engines in minimum 'burner. She was like a wild thoroughbred trying to break loose from the rope which tied her to the barn. Once those burners were lit, she expected to accelerate and didn't like being tethered to that boom. We were above 30,000 feet, though, and without the burners, at that weight and speed, we wouldn't be at thirty grand long. Somehow, after long minutes which pass agonizingly slowly, the miracle in the sky called aerial refueling was once again accomplished.

At this point I am anxious to start the climb and acceleration since every second now takes us closer to the threat area. The numerous disconnects have extended our end/AR point and I want to get this beast up to speed and altitude in plenty of time to enter the threat area.

First Walter must decode a priority message from the tanker. I fly loose formation on the KC-10 while Walter wades through his comm kit. We are heavy, we are low, and we are slow, and we are flying formation with a target. (There are two types of planes in the world; fighters and targets.) Worse, we are burning precious fuel at an astronomical rate and not climbing. The jet likes this even less than being tied up. Of course, I also realize that the message Walt is decoding may very well be a recall and we could be at the end of our mission anyway. I hope not.

Walter finally tells me he has the message and in his voice I note a certain degree of sarcasm seldom heard from the seat behind me. He informs me that higher headquarters wanted to let us know that the threat in the target area is extremely high, and have a nice day. I knew Walt was paraphrasing, but he told me that was pretty much the message. Speechless, I stroked the 'burners and started to climb.

Somewhere below us, Sardinian fishermen must have heard the sonic boom of our acceleration, but they probably cared very little about a large black aircraft speeding eastward.

Accelerating eastbound.

From the Rear Seat . . .

Day Three
0820

Once we got hooked up to the large KC-10, we kept getting a disconnect from the boom. At first, Brian thought it was his fault, but it kept happening and we felt there might be a problem with the boom latching device, since weather and turbulence were not factors. Finally, as Brian was reaching a fair amount of frustration, and we were receiving no gas, I figured out what was happening.

The tanker crew was trying to pass on a message to us via the boom intercom system which only transmits between tanker and receiver. Every time they keyed the mic button, it electrically disconnected the nozzle from our receptacle. Had it been our usual '135Q, equipped with special equipment which interfaced with the SR-71, this wouldn't have happened, but the equipment in

Subsonic.

the '10 was slightly different, and now was no time for me to explain it all to them over a secure frequency.

I tried one thing and luckily it worked. I turned down every radio we had and tuned in the KC-10 with absolute minimal volume. Brian hooked up and stayed connected. I told the tanker crew to maintain radio silence until we had received the gas and then transmit the message.

Brian and I felt the message might say to return home due to increased threat conditions. Now that we were this far into the mission, we really didn't want to turn back. I got my handy comm kit out to decipher the message.

At two minutes past the end/AR point, I received the message which told us, the threat to our aircraft in the target area had greatly increased and we could expect heightened hostile activity once near the Libyan border, and . . . good luck. I kept waiting for more, but nothing — that was it. *Good luck.* Now if ever there was a useless message, this was it. I know someone thought it necessary to tell us that, but since we were going in anyway, was this supposed to make us feel better? If there was some action we were supposed to do that was now different, it would have made sense, but, no, just a simple *thought we'd let you guys know your chances of getting hosed today are at an all time high* kind of thing.

Well, one thing that message did was definitely relax Brian and me, both. We understood the threat well, and respected it, but on this day, we feared nothing Qaddafi could send our way. Our

The view from 79,000 feet.

extreme faith in ourselves, and in the jet, was never stronger as a crew than at that moment. In the face of that message, it had to be.

Any mission tension we had been feeling up to that point now evaporated amidst inter-cockpit comments concerning a certain priority message. Brian was second to none in the sarcasm department and he did make me laugh. I don't remember us ever laughing like that on any other mission in the SR-71. It was a good thing to do at nine in the morning on this day. Brian lit the 'burners, and we got down to the serious business of getting the "take."

"... insects have splattered on the windshield of SR-71 aircraft at 70,000 feet. The existence of insects at that altitude was not explained, and it was not regarded as a flight hazard."

FROM AVIATION WEEK & SPACE TECHNOLOGY, OCT 1967

THE CRITICAL FORCE . . .

    My job took on many different names over the years and I ended up doing a variety of tasks, but basically I was a Lockheed field service rep. Before that, in the early '60s, when the first A-11 was still being assembled at the plant, I came to work for Lockheed in their experimental research department. Everything was very secretive in that you would know very little, if anything, about what was going on outside your own area, but all of us there knew that we were on the cutting edge of development at that time. Slowly, we started seeing individual pieces of the plane coming together. The landing gear, the tail structure, the large engine nacelles; it was just incredible seeing this plane take shape. I actually helped put the first bit of hydraulic fluid into the veins of this beast. It was not uncommon for us to work fifteen-hour days.

    Most of the men I worked with were older guys who had been with the Skunk Works for some time. Many of them had started with Kelly Johnson in the P-80 days. This was a marvelous collection of very competent gentlemen. There were only a few young guys like myself and I considered it a golden opportunity just to be there. Sometimes Kelly himself would come into our area with a small part in his hand, and he would tell us to break it. He wanted us to run it until total failure, to verify either his trust or distrust in a certain material. Remember, during this time we were still inventing the technology with which the airplane would be built. Up to that time, no one knew much about titanium, so many tests were continuously being run. We would run vibration tests, for example for millions of cycles to find the point at which cracks would develop.

    Kelly routinely was seen escorting any number of important looking dignitaries through the plant. One time, in the hydraulic plumbing area, I saw Kelly stop to actually forge one of the parts by himself. He always felt it was important that everyone knew that he knew what he was doing. He represented the absolute essence of the Skunk Works and the main reason it was given so much

respect. He was absolutely a fantastic person. I can't remember hearing a bad word about him from anyone. Not ever.

One time, a couple of his really experienced engineers made an erroneous estimation of some test parameters. It cost the program some materials and some time. When Kelly was notified of the error, and you knew he would be, he politely pointed out to the two engineers where they had made their mistake and then informed them that they were fired. Everyone was a little shocked over this, since these were two of his top engineers. The next day, Kelly told them they were hired again, and that he just wanted to make sure everyone understood that mistakes were acceptable, but bad mistakes wouldn't be tolerated. Kelly picked his people so well that I don't remember anyone ever really getting fired like that. One of the things I admired most about Kelly was his ability to solve problems with good old fashioned common sense.

After four years of working strange hours at the plant in Burbank, my wife was beginning to think I was the building janitor. I finally had an opportunity to head up to the ranch as part of a maintenance flight crew. This was very rewarding, getting an opportunity to see the plane actually going through its flight tests.

We would load it, in sections, on large flatbed trucks and in the dead of night would convoy into the remote reaches of the Nevada desert. Once we got to Groom Lake, there were crews there that could reassemble the plane, a difficult and interesting process in itself.

Interview of Kelly Johnson, who rarely gave interviews:

Kelly: *This is not going to be a long interview, is it?*

Reporter: *No, just a few short questions.*

Kelly: *Make them very short.*

Reporter: *What makes a good airplane design?*

Kelly: *There is no short answer to that.*

Reporter: *How important is an airplane's appearance?*

Kelly: *It is important to me. My wife tells me that if it doesn't look good, it won't fly good. That's good enough for me.*

Seeing the plane go from the construction stage to actually flying was truly motivational for all of us. We couldn't all help but feel very proud of our association with it while watching that beautiful black airplane go roaring off the desert floor. I'm sure we lost some of our hearing in those days, but we really didn't care; we loved seeing it and hearing it. That profile in the sky was so distinctive, right from the very beginning, and to me it remained that way throughout its entire life.

There was always an F-104 chase plane with it and we flew the Blackbird right in the middle of the day, with takeoff times deconflicted for the passage of overhead Soviet satellites. At certain times during the day, we would throw large camouflage netting over the planes and equipment, because we knew that was when the "other guys'" satellites were overhead. This was a thrilling time for all of us. We weren't all briefed on the aircraft's capabilities, but I knew from the intensity of heat on the landing gear after shutdown that high Mach was the norm.

Many problems had to be solved to fly an aircraft safely and consistently at high Mach, and none was a bigger headache than the inlet system. Initially we were having a lot of wiring difficulties. The heat was causing the wiring to crack, thus giving erroneous electrical signals, which in turn would misposition the inlet doors, causing havoc in the cockpit as the shock wave would be expelled from the inlet — a very complex system, one that was continually being worked on for the next twenty years, but the very heart of what enabled this jet to cruise at Mach 3 and beyond.

The constant flexing and expansion of the aircraft skin in flight was another phenomenon that had to be addressed. Fuel would leak into small grooves created by the expansion, then would be heated and cause some damage to the skin. The leading edge of the wing was originally a sort

Hustling during engine start.

of plastic material, and we finally had to cover it with the metal pie-plates you see today. New slip joints had to be installed once it had been determined where expansion was causing leaks.

Finding an adequate sealant for the fuel tanks was also a dilemma which lasted for the lifetime of the plane. We were asking the tanks to withstand temperatures of -40°F to 450°F. Because of the low volatile nature of the fuel, we accepted that the aircraft was going to leak. The Air Force never really took to this concept. They were used to their planes sealing a little tighter.

After nearly five years of experience with the plane, I had really fallen in love with it and decided that I wanted to go with it when it was delivered to the Air Force as the SR-71. I knew that with my experience on the A-11 and the YF-12, and my work in the experimental lab, I could bring a good understanding of the system with me to Beale. It was also a chance to work for Paul Mellinger, so I was happy when I was given the opportunity to follow the jet to Beale. I told my wife there was an opening for an experienced janitor in northern California, and she just laughed and started packing. She was always so supportive and never complained throughout everything.

At Beale I was responsible for helping the crew chiefs with the care and feeding of four SR-71s. In those days, the crew chiefs were not young kids but more senior master sergeants. These guys were master mechanics and a real joy to work with. They took a very personal pride in their work, and the atmosphere was definitely something very special. We were always looking for ways to improve things.

It was interesting working with the pilots too. They were all new to the aircraft, so sometimes their knowledge of the systems was a little sparse. They were all excellent aviators, but sometimes during the post-mission debrief, we had to try and interpret exactly what they were trying to tell us was wrong with the plane. What felt like a flutter to them was really a pulse to us. The MRS tape was invaluable for giving us the necessary data of what really happened in flight.

We all grew together in our understanding of the often mysterious ways of this unique plane. Unfortunately, just about the time the crews were getting very proficient, they would move out of the cockpit and up the ranks.

It always amazes me how lucky we were in not having more major accidents, which I guess is a tribute to the maintenance and the men flying the plane. I remember one time a jet taxied into the hangar following a flight, and as the pilot was going through the routine postflight control checks, the stick froze on him. It just wouldn't move, and he shut the aircraft down. When we opened it up to take a look, we found that the main push rod to the stick had come loose and disconnected right at the moment the pilot had taxied into the hangar. This was a very uncommon situation. Apparently, the locking device to the rod had not been installed properly, and all the vibrations of subsequent flights had finally worked it loose, luckily failing just minutes after the plane had landed.

When we lost our first jet at Beale, it was a little like losing one's child. We had come to know these first few planes quite well and regarded them as national assets. Everyone in those days really had their heart in their work.

I remember getting a call at home at four in the morning to come out to the base. When I got there at about 0430, I was directed to one of the shelters, and there sat an SR-71 with one main landing gear collapsed. Maintenance had been running gear retraction tests, and someone forgot to "safe" the left main gear. It was a sickening sight to see part of the wing crumpled on the floor. The maintenance folks felt really bad and didn't know what to do, so they called me. There wasn't much that I could do, but I was glad they called me. That's how we all felt; we were all in this together and we all felt a certain awe and respect for this plane.

In 1968 I was involved with helping set up the operation at Kadena. It was a different environment, as now we all not only worked together but were billeted together and often ate together. We became very close friends, and even though there was little time off, we didn't mind. It was an important mission we were performing there.

Our supply system was phenomenal. We could get a part in one day sometimes. Our priority was extremely high, and the logistics folks back home really supported us well.

The corrosion from the nearby salt water was not as much of a problem as we had anticipated, but we did experience some problems with the liquid nitrogen. The aircraft's hydraulic fluid was conditioned with LN_2. If the LN_2 was not extremely pure, oxidation would occur, causing a coking of the flight control servos. This was a serious problem and one that was difficult to pinpoint. We finally realized that the nitrogen purity we were getting in Okinawa was "only" 99.95 percent pure. For this jet, we needed medical quality 99.99 percent pure nitrogen. We even developed a mandatory filtering system for our hydraulic servicing carts to ensure there was no oxygen or moisture getting into the LN_2. Those logistics guys were great about getting us the pure nitrogen.

The planes were regularly rotated back to Beale for phase maintenance, and I guess I sort of started a little tradition one day. On one of the inside panels, I wrote in chalk a short message to the maintenance folks at Beale — something about changing one of the servos. Well I guess it appeared to them to be the equivalent of the "wash me" messages humorously drawn into the dust of a dirty car. Anyway, when the next plane arrived, you can bet there were numerous messages chalked on the inside of the main bay panels. That tradition continued from then on. I don't know how many really useful messages were transmitted that way, but some sure were funny.

Whenever one of the planes had to make an emergency landing somewhere off the island, a recovery team was dispatched to fix it and get it home. I was the first civilian to go with a recovery team into Southeast Asia. Normally, due to the hostilities of the war, civilians did not fly out to these locations. But there I was on board a C-130, flying from Okinawa to Utapao Airbase, Thailand, where a sick SR-71 waited.

I guess I was pretty naive, because as we flew in pitch darkness somewhere near the Laotian border, I noticed out the window all these flash bulbs going off. I was politely told by the GI sitting next to me that those were muzzle flashes of people shooting at the sound of our plane, hoping to hit us. I figured maybe I should get hazardous duty pay for this, but of course I didn't, and I never did think to ask about that. The reward for guys like me was fixing that jet and getting it back to fly again, and being able to do it with minimal men, equipment, and time.

The plane in Thailand had lost a generator, and we were able to quickly fix it. Of course, the GIs had this thing about leaving the little black silhouette of the airplane spray painted anywhere they could get away with. I'm sure certain B-52 crew chiefs didn't appreciate removing little HABU insignias from the gear doors of their planes.

There were so many good military guys behind the scenes in this program who, I felt, got very little credit. When the jet was setting all the speed records later, in 1976, I was there to assist the maintenance crews. Of course we had a spare plane in case one had a problem. Well I saw those guys work well into the night to get that spare ready for a speed run, just in case it was needed. Sure enough, after one good run, the primary plane unexpectedly crumped the next day, and without skipping a beat, they rolled that spare out and set those records. Everyone remembers the fantastic numbers attained during those flights, but to this day my best memory of it all was the faces of those mechanics who beamed with pride when they heard the results. To them, it was their

The beast roars in the runup area.

The chief closes the canopy.

Over the runway at Beale.

airplane, and they put a great deal of love and sweat into it. It's hard to describe just how much of a family people in the program were.

There weren't that many SRs in the world, and we often kept up with missions going on far from us. I was in Kadena in 1973 when the SR-71 staged out of the east coast of the U.S. during the Yom Kippur War and flew some remarkably long sorties over the Middle East and back. Everyone at Kadena followed those sorties with great interest and we all felt like we were right there with the crews, at least in spirit. These sort of things really helped the maintenance folks get into focus just how important their job was.

Sometimes it was easy to get complacent about working around the world's fastest jet. It was easy to forget just how special this plane was. Occasionally I would hear the newer enlisted guys in maintenance complaining about all the discrepancies which needed fixing before the next flight. I told them that may be true, but what other jet did they know that could cruise at three times the speed of sound with ease, and run in full afterburner for hours at a time with no restrictions, and go the places this plane could? Sometimes we had to keep our perspective.

I eventually came back to Beale and was fortunate enough to go on several very rewarding trips with the airplane. In 1974 Paul Mellinger did me a great kindness by appointing me the Lockheed rep on the Farnborough Airshow trip. My son was in the Army and stationed in Germany at the

time. Paul knew that with our schedule, I would likely not get to see my son for quite a while, so he put me on this trip knowing that I could rendezvous with my boy while over there.

I have to confess to feeling pretty elite standing out by that plane at such a big show as Farnborough, especially after the crews had set a speed record on the way over to England. While the jet was on display, everyone wanted to talk with the crews, and I was very content to stand quietly in the background. The crews deserved all that attention as they had done so much, so silently, for so long. I was content in the knowledge that I had known this airplane intimately in a way that the flight crews could not know. I had seen this plane go from the first assembly line construction, to combat in Vietnam, to setting world speed records, all while overcoming a myriad of unique problems. Part of me was in that plane as it sat there, and my great joy was watching the faces of the very enthused crowd as they gazed wondrously at this remarkable piece of American technology.

As a field rep for Lockheed on this program, I was given a great deal of authority and responsibility. Basically, I was speaking for the company when I was out in the field, and they were depending on me to make decisions concerning the health and security of that plane. How many programs do you know where the on-site rep can write an engineering disposition on the spot which gives approval to implement a change, and flight approval, until a more permanent fix is implemented? We could do that. It was a great responsibility considering the nature of the system we were working with and its importance to our national security. Paul and I, and the other reps, agonized over many decisions, but we would make them and we got things done. We only wished the Air Force could operate that way sometimes.

We often said to each other that there would never be another program quite like this, and I guess we were prophetic as we look back. Mostly I feel a great deal of pride today. I was just one person, but I could make a difference in a program that made a big difference to a lot of people.

It was a pleasure working with professionals like Paul Mellinger, the greatest diplomat I ever saw, often standing alone between Lockheed and the Air Force. We were involved with a one-of-a-kind program and continually seemed to be breaking new ground.

To me that airplane was truly beautiful. I never worked on any other Lockheed airplane, and I guess I never really wanted to. In all those years, I never tired of seeing that magnificent profile in the sky.

I am not ashamed to admit that I cried when the program was cancelled. I felt like I was seeing a friend and colleague forced to retire. An exciting chapter of my life was closed. There was so much life left in her; that's what made it so hard. If they ever made that phone call that said they were reactivating her, I know I would be out there to help.

I am happy in my retirement today, with the one regret that I could never fly the plane. Just once I would like to have experienced that incredible speed, and maybe an unstart too. I can remember vividly, watching the generals and the VIPs who were about to receive their one flight getting out of the PSD van and approaching the aircraft, and they would come face to face with this beast that now seemed to be alive, as it bled fuel and seemed to exhale whispy clouds of liquid nitrogen; and as they stared at the steps of the ladder to the cockpit, there was this definite look of fear and apprehension on their faces . . . and always in the background I would be saying, "I'll go, I'll go."

— Lew Williams —
Tech Rep
30 years

Lew Williams, in love with a jet for 30 years.

DINOSAUR ON PATROL

From the Front Seat...

D̄AY THREE
0920

Even at the high speeds we enjoyed, it surprised me how long it took to reach our turn point which would take us inbound toward Libya. While paralleling the north African coast, I began to get a feel for the enormity of this continent. Algeria seemed to go on forever. Watching my map display move across the small screen between my knees, I couldn't help but think of the airmen who traversed these waters during World War II. I was now getting a greater appreciation for the length of some of those B-24 sorties.

As we passed Sicily, the waters of the Mediterranean seemed to become even more beautiful, reflecting the sky with various shades of blue. I noticed several more airliners, though well below us this time. Commerce ships were trailing white water at random angles in all quadrants. Far to the north I could see the snow-capped Alps of northern Italy. That we were about to enter an area of intense hostile threats seemed quite incongruent with the picture before me. I looked to my right and knew the coast I now saw in the distance was Libya. I concentrated my eyes on the instruments before me.

Early in the accel, Walt and I both knew that the outside air temps were worse than forecasted, as the jet accelerated and climbed at a slower pace than normal. The inlet door continued to buzz and wasn't helping our gas situation at all. Walter was concerned that we wouldn't reach our programmed altitude and speed as we were now behind the profile.

I reassure him that the jet will make it. I feel it will. I'm not sure there was one particular thing which made me feel good about 960 right then, but I did. I liked what I was seeing in the cockpit; the spikes were moving dead on the money in total unison, the engine temperatures were stabilized within a couple degrees of each other, and the variable engine nozzles were stable. Then, too, there were the intangibles. The jet, aside from some of the bad sounds the inlet was producing, just felt solid.

When I was a young lieutenant with little flying experience, I remember flying with a crusty old major who was my instructor at the time. He would tell me that it wasn't enough to fly the airplane, you had to feel the airplane. He also said that once you learned to feel the plane, you could get it to really work for you, as it then became an extension of yourself. At the time, I thought he was pretty much of a dinosaur, but I never met too many pilots who flew with as much finesse.

As I found myself talking to 960 and feeling her every pulse, gently nursing her altitude higher and telling Walt with certainty the jet would do fine, when in fact the numbers looked dismal, I realized something. Somewhere between the white-knuckled death grip on the yoke of that first Cessna, and the gentle caress of a golden space suit glove across the auto-pitch wheel of the world's fastest jet, I had become that crusty old major. In the blur of the past 3,000 flying hours I had repeated his words many times to my own students. Now, flying a nearly 30-year-old jet, and loving every minute of it, and talking to her, and feeling like I was sitting in a classic piece of aviation construction, the likes of which we'd never see again, I was the dinosaur.

As we crest 65,000 feet, I notice the right nozzle go full closed. I immediately check the EGT and the RPM on that side. They remain stable. If the nozzle has really closed as indicated, I would be seeing higher temps and erratic RPM. I suspect the gauge has failed. I also check the turn and slip indicator for any yaw input that would have resulted with one nozzle closed and the other open to the proper position. I hold my breath for a couple seconds and then feel secure in my analysis that the nozzles were fine and that the gauge has failed. The jet seems to be toying with me. We had been given this exact scenario in the sim, several years ago, and I never forgot it. I believed the gauge in the sim that day and failed to verify the condition through a series of cross-checks. Walt and I performed an emergency landing that time, with nothing more than a simple gauge failure. I am glad now that we had that one in training. I don't bother Walt with the concerns of the front seat now, as he is busy working sensor checks, and we press on with a buzzing inlet door, one bad gauge and two good engines.

The intensity in each of our cockpits is boldly interrupted by an unexpected radio transmission. Somewhere in the blue waters below us, the U.S. Navy Fleet is tracking us and just wants to let us know they are with us in spirit. Someone in the carrier task force transmits, "You guys have a good one . . . ." — very unexpected but highly appreciated. My thumb twitches involuntarily across the mic button on the throttle, and I have to fight the urge to reply.

Walter informs me that the DEF has passed all self-tests and is good. This is an abnormal call by Walt, but I know that he is doing it to help ease any concerns I may be having about the SAM threat. It helps.

Two thousand pounds below programmed fuel, we start our turn toward the Gulf of Sidra and the ANS smoothly rolls 960 out of the turn, pointed squarely at Qaddafi's "line of death."

From the Rear Seat . . .

Day Three
0930

As we accelerated into the Med, it soon became apparent that the combination of hot outside air temps and the loose forward inlet door were greatly affecting the jet's performance. We were already 1,500 pounds below programmed fuel. I had seen fuel figures like this before, and Brian seemed to somehow always find a way to make it up en route, so I never got too worried until we were about 2,000 pounds down. But of more concern to me was the climb profile. Due to the threat, I definitely wanted us to be high and fast when we entered the target area. I knew, too, that with a semi-faulty inlet door, we would be better off getting to our programmed Mach early enough to see if it was going to hold steady with the different atmospheric conditions we would experience. The temps were unbelievably bad and I could only hope that they would improve as we slugged along through 1,400 knots in the climb.

As we made our way past Sardinia and then, shortly thereafter, Sicily, I was able to use the viewsight to confirm our position with the ANS and again it showed us dead on course. I couldn't help but think what a pretty day it was as I looked down on the islands in the sea. At this point, I knew we were a go, since even if the main camera failed completely, we had two good backup cameras functioning fine. My main concern was that the jet would perform perfectly. We often asked that of the jet, due to the nature of some of our routes, and we were asking it again this day.

I asked Brian how it looked in the front and if he thought we would make the necessary altitudes and speeds. In the midst of cursing the inlet door, he assured me that the jet would make it fine and that it felt good to him. Based on the Mach and altitude I read on my TDI and the time to go until we start our turn toward Libya, I had no reason to believe him; except that I had flown with this man long enough to know that when he said the jet felt good, though I had no scientific data to understand this statement fully, I believed him. He never told me how to work the ANS and I never told him how to fly the jet. I really hoped he was right this time.

I couldn't spend too much time worrying about the climb profile, though, as I was still trying to recover the OBC. I tried everything I knew to do, and it looked like it might work. I then re-checked the aperture settings, sun angle settings and counters on the backup TEOC cameras, as I knew they very well could be the primary picture takers that day. This was fine since, in talking with the tech reps at Mildenhall, I had learned that the TEOCs usually produced the highest resolution pictures due to their superior stabilizing gyros.

Finally, everything was about as set as I could make it in the back seat, and it was then just a roulette game watching the Mach and altitude creep up through a hotter than normal stratosphere, waiting to see if we would make it at least to 3.2 and 74,000 feet. We were still below the desired fuel, close to 2,000 pounds now and I became more conscious than ever of the buzz from a partially opened forward door. At that speed, the slightest spillage of air overboard can create significant drag, not to mention unwanted vibrations for the sensors. We lived and died with every hundredth, and every tenth, of Mach number as the counters on the TDI crept higher.

I switched the backup INS over to Akrotiri Air Base in Crete, as our position now dictated it would be our best alternate in case we needed to come down. This meant that in a few moments we would be starting our turn, inbound, to the very large continent to our right.

Then something happened which startled us both. Transmitting loud and clear, on the primary strike frequency, which was not to be used unless absolutely necessary, was the U.S. Navy. We had not heard a thing all day on frequency, and it really got our attention. Apparently, the Carrier Task Force which had been patrolling the Mediterranean for the past several weeks, was standing ready to offer any assistance to our missions and was obviously aware of our position. In a clear, calm voice, someone transmitted to us, "You boys have a good one. Go get 'em." No decoding was necessary. We issued no reply, though I know, deep down, Brian really wanted to. That radio silence had been broken was less important to me at that moment than knowing that the fleet was out there to back us up if necessary.

We started a right turn and in the viewsight I picked up the coastline of Libya.

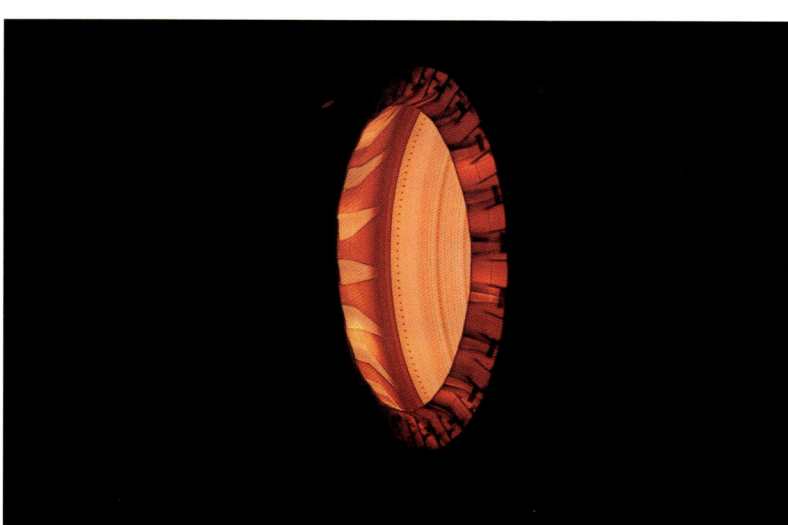

Engine nozzles glowing from heat of J-58.

THE CRITICAL FORCE...

The J-58 was unique. There was no other jet engine made like that, ever. It was definitely the heaviest engine I ever saw on a jet. Everything about it was big and heavy, and very tough throughout. One thing different on it was all the plumbing was attached externally. This made it look even bigger. I have torn that engine down as far as it would go, into thousands of pieces and really felt it was the best engine for what we were asking it to do. After a while, I could just listen to that thing running and tell if something wasn't just right.

I worked with that engine from 1963 until 1988. I guess you could say I was an "old head" after a while. A lot of people just assumed that I worked for Lockheed or Pratt and Whitney, but actually the company that hired me was Aviall, and that's the only company I worked for. They were contracted out to take care of that engine, and I guess you could say I was one of the caretakers. We were kind of on our own out there, as we never saw anyone from the home office, except for once a year a guy would come to the DET to have us re-sign our contract. Pratt and Whitney never said anything to us, so I guess we were doing OK. If we weren't, I know we would have heard a lot about it.

I saw the jet for the first time out at the ranch. I looked into the hangar and didn't know what in the world it was. I was pretty certain that it couldn't fly; it was just too big and weird shaped. Whatever it was, we were certainly briefed thoroughly before flying out to the ranch. We got the

Always looking for simultaneous 'burner lights on takeoff.

J-58. Lots of external plumbing.

Doc carefully observing engine run.

impression that if we leaked word of the jet, they wouldn't just cut our tongues out but cut our heads off too. My wife had no idea what I was working with for about five years.

About a year after I had first seen the plane, I got to see it fly, again out at the ranch. Well, I sure changed my mind about the aerodynamics of the thing. It was beautiful in the air. We had no idea where it was flying at the time and knew not to ask. My area of concern was the powerplant and I proceeded to learn everything I could about that engine.

Early on, there were some growing pains when the J-58 was mated to the SR-71. Like most systems on the jet, heat was the major factor in causing problems. The first fuel controls we used just weren't standing up to the high temperatures. We initially tried to build a little cocoon device to keep it cooler, but eventually we had to go to a newer Bendix system and that worked much better.

We were seeing lots of stress cracks in the afterburner liner initially, and someone came up with this terrific ceramic material that was then used. This material could withstand immense heat and then, within minutes after engine shutdown, it would be cool to the touch.

One day a group of scientists and engineers were touring the facility and they were particularly impressed with the properties of the ceramic lining. I was just there to make sure no one hurt that engine and answer questions if necessary. Well, there I see one of these guys scraping the lining with his fingernail, trying to get some particles on a handkerchief, probably to take back to his company to study this material. All he got was a bruised fingernail.

That engine was one tough machine. People were very concerned about the amount of flight time spent in full afterburner, but this didn't hurt it at all; in fact, I think it was good for it. It ran better in 'burner, and I just think that it made it tougher. You could loosen fillings in your teeth standing close to that thing in full blower.

Rare tail art chalked in by crew chief.

When we were doing an engine run we would listen for sounds out of the ordinary, and after a while you could hear little things that clued you to potential problems. We also watched the flame pattern. You could tell if the fuel control was working properly by looking for changes of flame color. If we saw light and dark patterns in the 'burner flame, we knew something was wrong. We never hurt one on the test stand though. That was our baby, and there weren't that many of them around. I never had one blow up on the test stand, but I had a couple that sounded like they were going to.

There were no left and right engines; they were completely interchangeable, and they burned very clean. There were only two fluids in that engine — oil and fuel. Of course the fuel also served as a hydraulic fluid. Right after engine shutdown, you could crawl through that engine with a white shirt on and come out without a mark on you. Now you sure couldn't do that with, say, an F-4. The airplane leaked but our engine didn't.

When I was assigned to DET 4, I enjoyed it more since we were closer to the mission and everyone there really worked together well. I usually didn't know much about the routes of flight, but in the debriefings I would learn a lot about where the crews had flown.

No one wanted to have a mission cancelled due to any problem in their system, and we were no different. When we would watch the crew do their engine runup on the runway just prior to takeoff, our hearts would sink if we heard the throttle coming back, which usually meant

Brute force.

something was wrong. We always listened to the 'burners lighting off too, as we wanted those engines to light off at the same time, which I know wasn't always the case.

Many times the engine was blamed for an abort, but upon further inspection, the inlet was found to be the culprit. That big spike could be slightly loose on its rollers and at high speed would send a vibration throughout the plane that surely felt like the engine was acting up. There was a very distinct division between inlet controls and the engine. Just remember, you had a whole lot more inlet unstarts than you had engine flameouts, and the fact that during all those horrendous unstarts, that engine kept right on cooking says a lot. Why, that thing could digest more FOD than any engine I've ever seen and just chew it up. We had one come back one day that must have sucked up a rock or something, and it had torn several turbine blades in two, and it never skipped a beat.

About the only problem we never really fixed was the spray nozzles in the 'burner section. They kept burning up and then we would change them, and this caused problems because we never knew if we were sealing them tight enough to keep fuel from leaking into the 'burner section. They never did develop a better seal for us. I saw one leak on the test stand once, in full afterburner, and it threw flames in about six different directions.

Now that I have learned much more about the actual missions, I am amazed at just how far away from the home base the crews would fly. That must have taken a great deal of coordination. I know we spent some long nights in the hangar during the Libyan crisis. Basically we didn't go home; we pretty much just hung out at our shops for the whole three days. The launch times were very close, with two jets going each day for three days. That put the pressure on everyone. We didn't want the mission to be cancelled because of an engine problem.

To me, the pilots were special folks. They had to go out and validate that engine every time they flew and test its performance many miles from a safe landing site. If a crew was late returning from a mission, we all felt it and were nervous until we saw that big bird on final approach. I'd usually get my best debrief from the pilot right at the bottom of the ladder after a flight, and before the crew was even out of their suits, we would be crawling over that engine to check it out.

Looking back, I believe the program lasted as long as it did because there were so many dedicated people involved with it. I was very proud of my part, but I feel that I contributed, really, very little compared to so many others.

To this day, I feel there are many people who have no idea what a technological marvel this aircraft was for its day. I always wondered just what it must have felt like sitting up there in full afterburner, feeling that engine, and sustaining that amazing speed . . . .

— *Doc Strange* —
J-58 Field Rep
25 years

"... in 1958 I was already thinking of something like the SR-71. Never before had anyone designed an aircraft to maintain supersonic cruise. From the time we started work until the first flight took twenty-seven months, and nothing came easy on this plane. The normal aluminum finish we had used on other planes was good to about 250°F, but we were now asking the skin of the plane to endure an average temperature of 550°F. We literally had to invent a hydraulic fluid that would operate at 650°F. We went through seven different types of oil pressure transducers to find one that withstood 1,500°F air. Constructing the nose section was the most difficult since it would require the thinnest skin, would run the hottest in flight, except for the engines, and required the fitting of a windscreen. On top of this, we had to figure a way to pump plenty of cool air up there to keep the pilot from melting.... The first 6,000 parts we made all had to be scrapped. They had become too brittle and some welded joints had separated. The chlorine in Burbank's water system had poisoned the titanium parts, so we had to invent a cutting fluid in order to shape the titanium. I finally posted a reward of $50 to anyone at the plant who could find something on this plane which could be accomplished easily. I've still got that $50."

KELLY JOHNSON

THE CRITICAL FORCE...

As an Air Force sergeant, my area of expertise was electronic warfare and I was initially assigned to the U-2. The first time I saw the SR-71 I couldn't believe that it could fly as fast as it did with such an unusual shape. For me, it was truly an invigorating experience to be right out by the runway when the aircraft took off. I think the maintenance guys enjoyed watching that as much as the pilots did.

I received my checkout with the SR-71 DEF systems from the civilian tech reps there at Beale. They were the experts, and their incredible knowledge and patience made my training a most enjoyable experience. They were right there alongside us every step of the way. The tech data publications were very inadequate at first, especially for us maintenance guys who relied on them. I offered to initiate and rewrite several checklists on the spot. Much to my surprise I was given an immediate OK to do so. I knew it didn't always work this way in the military system, but this program was so unique in so many ways that the unusual simply became commonplace. I guess I got spoiled. The tech reps assisted me greatly with this project and we were able to turn some of the engineering prints into workable checklists for the Air Force. There was, of course, lots of trial and error involved.

Electronic defensive systems became more important to the Air Force as the threat from Soviet-made surface-to-air-missiles in the Vietnam War became more potent. In the early '70s I was dispatched to Okinawa where we had set up our first permanent detachment for the SR-71. I ended up staying there eight years.

Our biggest problem with the DEF system was basically trying to keep the system cool in flight. The DEF unit puts out a lot of power, and if aircraft cooling failed, it would burn up. When that happened, the RSO would get a warning light and have to shut the system down.

The several inches of growth that the aircraft experienced at high Mach caused us some problems with connectors coming loose in the bay which housed the DEF. We always double-checked these things prior to flight, but the unit barely fit into the bay as it was, and it made it even more difficult to get to anything once installed. The units changed over the years to keep up with changing anti-aircraft weaponry, and we did see some improvements in DEF pod design.

Just preflighting that thing was an arduous task, as each half of the system weighed 400 pounds. There were numerous electrical checks necessary. Unlike the ANS, or some sensors which were always loaded in the same place in the aircraft, our systems would be shifted around depending on the type of sensors being loaded on that particular mission.

The real bear for us, though, was when one jet broke, and the spare jet needed to be readied in minimum time. Downloading and uploading again always increased the possibilities of some component not interfacing just right. And I will tell you, each and every one of those jets seemed to be made just a little bit different. The units just fit better into some airframes than others.

Keep in mind, too, that we weren't just loading defensive systems but also electronic intelligence systems which were designed to detect ground sites that could be a threat to the aircraft. Some of the best and most important data this aircraft ever gathered was on the ELINT tapes.

As maintenance guys, we were highly enthused about the missions this aircraft was flying. We usually had a pretty good idea where the crews would be flying and would set the DEF systems accordingly. Most of the electronic activity that we saw came from Korea. When they fired that missile in the early '80s at one of our planes, we were very proud of the way the DEF systems responded.

We were not simply training for a bigger mission someday; we were actually performing the prime mission of the aircraft every day. Getting ready for those missions was always a little hectic. We were normally given a 24-hour notice to ready the aircraft. Oddly enough, our equipment took five different bays to house it all and needed more space than all of the cameras. It was necessary for us to coordinate loading with the other systems on board. For example, if the TEOC Cameras were going that day, we would load first, since there was very little clearance between our pod and the cameras, and we didn't want to risk scratching the gold plated lens on the camera with our cables.

Then of course, there was the constant leaking of fuel, which not only made it slippery on the hangar floor but also caused us some problems when fuel would seep into the bays where our pods were and coat them inside and out.

Our first indication of any problems with our system during a mission was listening to the RSO at the bottom of the ladder after a mission. It was pretty much an on-off thing for him in the cockpit. All the programming of the system was done on the ground prior to the flight.

The quick debrief at the bottom of the ladder.

I felt, overall, it was a very capable system. Over the years we went from the A model to the H model, and that H model was an extremely capable system. One time, while still using the DEF A, we had to swap a broken system out and had very little time to exchange units in the jet. In order to avoid an aborted mission we ended up hand carrying the 128 pound pod across the hangar and installing it in time, safely. It was definitely not by the book, but the mission went on time.

I really enjoyed that we could talk to the crews too, one on one. We had a great deal of respect for the officers in those space suits. They truly seemed to enjoy flying that plane. We always felt like they were risking their lives out there. Sometimes new enemy weapons would come on line, and the guys would be tasked to fly certain routes, hoping to activate that site in order to record the electronic signals. On one flight I remember they brought back more data on a new Soviet SAM system than SAC had been able to acquire in over two years. That had to be a little scary for the fliers.

You couldn't work with this program as long as I did and not have at least one adventure story. Mine occurred when one of the jets on a mission from Kadena had a problem up north and was forced to land at one of our Air Force bases in South Korea. I was sent from Okinawa as part of the recovery team. The plane was fixed rapidly, and the following day, we were on the field to watch the takeoff.

Well, about halfway into the takeoff roll, the nosewheel assembly of the jet goes hardover full left. The jet didn't much care at that point, and simply continued straight down the runway, burning nose tires as they were skipping ninety degrees off center down the runway. The pilot later said that the vibration from the skipping nosewheel assembly made it impossible to read any gauges and he felt like a piece of popcorn in a popper.

The force in action.

Well, it must have been pretty severe, because as we watched the cocked nose strut, now minus tires and melting metal against the runway, I notice to my horror that a bay door had been vibrated open and out pops the DEF F pod. We now have two moving objects on the runway — a decelerating SR-71 and a 635-pound pod bouncing alongside, eventually passing the jet. I believe we all actually flinched with each bounce. Well our first concern was with the crew, and they were able to finally stop the jet, which was now sitting off the side of the runway with both engines on fire.

We all piled into the launch truck and sped right on out there at about 120 miles per hour. I remember thinking at the time that this would surely be the accident of the day, ten guys hanging on in the back of the pickup at breakneck speed. Well the crew had egressed the plane and were all right. The fire was extinguished and we finally located our pod.

There it sat in the mud. Not a pretty picture. The fasteners which held that pod in place were extremely strong, but all of them had been smashed. Everything in that bay had been beat to death from the vibrations. When we finally got the jet off the runway, we were not looking forward to opening the other bays, afraid of what we might find, or what might fall out on top of us. We had a minimum crew, and with no DEF systems support equipment, we slowly opened each bay one at a time.

We ended up spending thirty days there, until they could bring in two new engines and new nose tires for the aircraft. The plane then flew back to Kadena uneventfully. The thing that impressed me the most after it was all over, besides the fact that we weren't all killed in that truck ride, was how incredibly strong the jet was. That nose gear strut never collapsed under all that stress.

Eventually I had to return from Okinawa and was at Beale until the very end of the program. We really couldn't believe that the program could be cancelled in light of the tremendous data we were retrieving from the sensors. I believe that SAC had a fairly lackluster approach to the plane and didn't really understand all of its capabilities. That attitude really helped put the jet in its coffin.

There were many secondary missions going on electronically each time it flew, which I don't want to get into, but suffice it to say that the Navy and the Army now sorely miss the product this jet delivered weekly.

I know that whatever else I do in my Air Force career, I will be most proud of my association with that plane and its mission. I really can't compare it to anything else.

To me, every mission was critical and I treated that DEF system like the crews' lives depended on it, because they very well could. With the thousands of missile launch indications received by the DEF over the many years this aircraft flew, I am very proud to say we never lost one jet to enemy fire.

— *Master Sgt. Mike Hull* —
Defensive Systems
14 years

The driver's seat.

ACROSS THE 'LINE OF DEATH'

From the Front Seat...

Day Three
0950

With the Libyan coast fast approaching now, Walt asks me for the third time if I think the jet will get to the speed and altitude we want in time. I tell him yes. I know he is concerned. He is dealing with the data; that's what engineers do, and I am glad he is. But I have my hands on the stick and throttles and can feel the heart of a thoroughbred, running now with the power and perfection she was designed to possess. I also talk to her. Like the combat veteran she is, the jet senses the target area and seems to prepare herself. For the first time in two days, the inlet door closes flush and all vibration is gone. We've become so used to the constant buzzing that the jet sounds quiet now in comparison. The Mach correspondingly increases slightly and the jet is flying in that confidently smooth and steady style we have so often seen at these speeds. We reach our target altitude and speed, with five miles to spare.

Entering the target area, in response to the jet's new-found vitality, Walt says, "That's amazing . . ." and with my left hand pushing two throttles farther forward, I think to myself that there is much they don't teach at engineering school.

Out my left window, Libya looks like one huge sandbox. A featureless brown terrain stretches all the way to the horizon. There is no sign of any activity. Then Walt tells me that he is getting lots of electronic signals, and they are not the friendly kind.

The jet is performing perfectly now, flying better than she has in weeks. She seems to know where she is. She likes the high Mach, as we penetrate deeper into Libyan airspace. Leaving the footprint of our sonic boom across Benghazi, I sit motionless, with stilled hands on throttles and the pitch control, my eyes glued to the gauges. Only the Mach indicator is moving, steadily increasing in hundredths, in a rhythmic consistency similar to the long distance runner who has caught his second wind and picked up the pace. The jet was made for this kind of performance and she wasn't about to let an errant inlet door make her miss the show. With the power of forty locomotives, we puncture the quiet African sky and continue farther south across a bleak landscape.

Walt continues to update me with numerous reactions he sees on the DEF panel. He is receiving missile tracking signals. With each mile we traverse, every two seconds, I become more uncomfortable driving deeper into this barren and hostile land.

I am glad the DEF panel is not in the front seat. It would be a big distraction now, seeing the lights flashing. In contrast, my cockpit is "quiet" as the jet purrs and relishes her new-found strength, continuing to slowly accelerate. The spikes are full aft now, tucked twenty-six inches deep into the nacelles. With all inlet doors tightly shut, at 3.24 Mach, the J-58s are more like ramjets now, gulping 100,000 cubic feet of air per second. We are a roaring express now, and as we roll through the enemy's backyard, I hope our speed continues to defeat the missile radars below.

We are approaching a turn, and this is good. It will only make it more difficult for any launched missile to solve the solution for hitting our aircraft. I push the speed up at Walt's request. The jet does not skip a beat, nothing fluctuates, and the cameras have a rock steady platform.

Walt receives missile launch signals. Before he can say anything else, my left hand instinctively moves the throttles yet farther forward. My eyes are glued to temperature gauges now, as I know the jet will willingly go to speeds that can harm her. The temps are relatively cool and from all the

warm temps we've encountered thus far, this surprises me . . . but then, it really doesn't surprise me. Mach 3.31 and Walt is quiet for the moment.

I move my gloved finger across the small silver wheel on the autopilot panel which controls the aircraft's pitch. With the deft feel known to Swiss watchmakers, surgeons, and "dinosaurs," I rotate the pitch wheel somewhere between one-sixteenth and one-eighth inch, locating a position which yields the 500-foot-per-minute climb I desire. The jet raises her nose one-sixth of a degree and knows I'll push her higher as she goes faster. The Mach continues to rise, but during this segment of our route, I am in no mood to pull throttles back.

Walt's voice pierces the quiet of my cockpit with the news of more missile launch signals. The gravity of Walter's voice tells me that he believes the signals to be a more valid threat than the others. Within seconds he tells me to "push it up" and I firmly press both throttles against their stops. For the next few seconds I will let the jet go as fast as she wants.

A final turn is coming up and we both know that if we can hit that turn at this speed, we most likely will defeat any missiles. We are not there yet, though, and I'm wondering if Walt will call for a defensive turn off our course. With no words spoken, I sense Walter is thinking in concert with me about maintaining our programmed course.

To keep from worrying, I glance outside, wondering if I'll be able to visually pick up a missile aimed at us. Odd are the thoughts that wander through one's mind in times like these. I found myself recalling the words of former SR-71 pilots who were fired upon while flying missions over North Vietnam. They said the few errant missile detonations they were able to observe from the cockpit looked like implosions rather than explosions. This was due to the great speed at which the jet was hurling away from the exploding missile. I see nothing outside except the endless expanse of a steel blue sky and the broad patch of tan earth far below.

I have only had my eyes out of the cockpit for seconds, but it seems like many minutes since I have last checked the gauges inside. Returning my attention inward, I glance first at the miles counter telling me how many more to go until we can start our turn. Then I note the Mach, and passing beyond 3.45, I realize that Walter and I have attained new personal records. The Mach continues to increase. The ride is incredibly smooth.

There seems to be a confirmed trust now, between me and the jet; she will not hesitate to deliver whatever speed we need, and I can count on no problems with the inlets. Walt and I are ultimately depending on the jet now — more so than normal — and she seems to know it. The cooler outside temperatures have awakened the spirit born into her years ago, when men dedicated to excellence took the time and care to build her well. With spikes and doors as tight as they can get we are racing against the time it could take a missile to reach our altitude. It is a race this jet will not let us lose. The Mach eases to 3.5 as we crest 80,000 feet. We are a bullet now — except faster.

We hit the turn, and I feel some relief as our nose swings away from a country we have seen quite enough of. Screaming past Tripoli, our phenomenal speed continues to rise, and the screaming Sled pummels the enemy one more time, laying down a parting sonic boom.

In seconds, we can see nothing but the expansive blue of the Mediterranean. I realize that I still have my left hand full-forward and we are continuing to rocket along in maximum afterburner. The TDI now shows us Mach numbers not only new to our experience but flat out scary. Walt says the DEF panel is now quiet and I know it is time to reduce our incredible speed. I pull the throttles to the min 'burner range and the jet still doesn't want to slow down. Normally, the Mach would be affected immediately when making such a large throttle movement. But for just a few moments, old 960 just sat out there at the high Mach she seemed to love and, like the proud Sled she was, only began to slow when we were well out of danger. I loved that jet.

Walt picks up some signals well out over the water and correctly interprets them to be Navy F-14s somewhere below us. I realize that I am sweating profusely in the suit and reach to turn my suit temperature down, only to realize that it has been at the cold setting for the past hour.

Beyond Mach 3.2. Like a bullet — except faster.

After landing, taxiing in.

It's been a long day for the other crew, too. Bernie goes through shutdown checks.

Because of our excessive speed, the descent point for our final refueling is fast upon us. We are no longer short of fuel, either, as our time above Mach 3 has actually made up some gas for us.

In just a matter of moments, the jet has carried us from the hostile frenzy of a battle zone to the calm blue waters of the Mediterranean, complete with the commercial ships and planes I had seen earlier, no doubt transporting tourists to some of the exquisite islands below us. This mercurial transition exceeds our immediate ability to digest it all and manifests itself in a feeling of slight euphoria between the cockpits.

By the time I eased up behind our tanker, we were feeling pretty good, and I don't think we ever had a better refueling. The KC-10 crew talked with us on the boom interphone and shared our joy over a successful run.

As we start the final leg home, the jet accelerates sluggishly through familiar hot outside air temperatures. The inlet door flutters and settles into the constant buzz that I am too tired to curse. Walter attempts to explain to me, very scientifically, that it was some sort of cool air mass across the Libyan coastline which enabled the jet to perform so well. I know better. Our jet vibrates all the way home, and I forgive her.

Mildenhall is covered with near-thunderstorm weather. For the first time in two days, I start to feel the fatigue. I shoot an instrument approach through turbulence and rain, and 5.7 hours after taking off, I crunch the plane down with an uncharacteristically hard landing. Walter is kind and does not say anything like he normally would. We are both beat, and it feels good to pop the gloves off as I taxi toward the hangar.

FROM THE REAR SEAT . . .

DAY THREE
0955

As we turned our nose toward the land mass of Libya, two things happened which I could not explain; the OBC started working normally and the vibration from the loose door stopped. This plane just seemed to know where it was and exactly what its mission was.

I had the DEF all set, and initially, not one peep from the bad guys. I hoped maybe we had surprised them. I took a quick glance at the TDI and noted that Brian had milked the jet up to the necessary Mach and altitude, and then some. (Later, he would credit the jet fully for this.)

Racing toward the first target area, everything looked good, and I had about one second of total calm before the DEF panel lit up like a Christmas tree. This told me we had surprised no one and that probably every camel driver in north Africa was tracking us with a variety of SAM radars — lots of them. For the first few moments there were no launch indications, just acquisition signals. Our great speed was giving them a slight problem and I knew it would probably be only a matter of seconds before a launch occurred. I also knew that the DEF pod we were carrying put out enough jamming voltage to light a small city, so I tried to think about that as I peered into the viewsight.

There was a slight trace of fog on the coast but otherwise it was clear, with no dust. The viewsight confirms to me that the cameras are looking at the right areas. As we hit the coastline and headed inland, the cameras were automatically sweeping left and right, then looking straight down to capture the areas of interest. It is beautiful to watch. The ANS shows zero error on course. The jet is smooth and stable for the first time all day. The cameras are clicking away, and although I have no way of knowing for sure if the OBC is working, I feel good about the TEOCs and I know that we are getting something on film.

Things then begin to get really busy for me. In between adjusting settings on the cameras, I hit some buttons on the DEF in response to the multiple threat signals that now number more than I have ever seen. I see the DEF responding repeatedly to hostile signals. The signals are now indicating launches on our aircraft. As the DEF is doing its thing, I peer into the viewsight and increase the magnification in an attempt to see any telltale signs of missiles coming up at us. I don't, and quickly bring my head up to monitor the next turn on course. The auto-nav responds perfectly and we are already one-third done with the entire target area.

At this point I get some very serious launch signals. The DEF responds and I alert Brian that I would like an increase in Mach and altitude — now. One thing about Brian, you never had to ask him twice for more Mach. The jet effortlessly rolled up the speed as Brian went to full 'burner and started a slow climb. We both knew that a hard turn away was coming up shortly and that would really be our best defense. The question was, could we wait that long, or should we turn now. We said very little and kept getting faster. I had a good faith that between the DEF and our speed, and the upcoming turn, they wouldn't hit us.

The turn was only 45 seconds away and it took about a day and a half to get to it. I activated another switch to ensure the aircraft would capture the missile engagements in its electronic sweeps. I knew Brian's attention was glued to those inlet doors and he knew not to talk much to me as I was heavily involved with the heart of the mission.

I know we both sighed with relief as the jet rolled into that final turn, away from the land mass. I glanced at the Mach and realized we were really hauling as it had continued to increase. It gave me a good sense of security seeing those numbers on the TDI. Apparently, we had encountered cooler temps over the land mass, and the jet really liked it. Going through 3.45 I realized we were setting our own personal highs for speed that day. I noted the true airspeed on the ANS above 1,900 knots and rising. The Mach continued to climb and somewhere past 3.5, I told Brian he could definitely come back a little on the power as the DEF panel had by now calmed down. We were speeding so fast out of there that we actually were ahead of the programmed fuel by the time we started down for the tankers off the coast of Algeria. Unlike other planes, the faster the SR-71 flew, the better the gas mileage.

Once on the boom, the tanker guys all wanted to chat on the boom mic about the mission. I guess it is pretty exciting for them too, being a vital part of the whole operation. I treat myself to a tube of butterscotch pudding, and it never tasted better.

On the last leg home, Brian asked me if I thought we got what we came for and I told him yes. We felt good about it. It's important for fighter guys to know that they accomplished the mission. We only spent about twelve minutes in the target area, but it seemed like about thirty. I knew our cameras were likely picking up the crash site of the F-111, if, in fact, it went down over land. I still wondered who the two guys were that had gone down during the attack.

On the way out of the Med I witnessed the beautiful sight of the Rock of Gibraltar in the viewsight. I thought to myself that this is probably the best view of this landmark I will ever have in my life. It was a magnificent view as we start the turn northward.

I take some notes on my knee pad concerning some things I saw on the DEF panel. The debriefers will be interested in everything we can tell them. I start to feel exhaustion now and turn my vent up to run some cool air through my suit. We have flown two consecutive days and I wonder how many more missions, if any, we might fly.

Based on our fuel, I calculate we will arrive at Mildenhall with enough gas for one approach, barely.

Somewhere along the route home I am, once again, aware of myself breathing in the space helmet. I don't remember hearing it at all for the past hour.

THE CRITICAL FORCE...

As soon as I got into the program with the Logistics Command, I found that I was immediately looked upon as the expert for whatever particular problem we were working. Of course I wasn't, but at times, the operators would look to DET 6 as if we were. We were somewhat of the middlemen between Lockheed engineering, spare parts and the Air Force.

Military commanders involved with daily operations of the aircraft wanted a lot of problems resolved, and therefore a lot of questions answered, immediately if not sooner. This was not an easy aircraft to maintain, and translating original engineering data to workable field maintenance was challenging, to say the least. It really was a learn-as-you-go program. I know many of the folks we dealt with assumed we all had big time engineering degrees, but in actuality most of us did not have a four-year degree at all. In other programs, the logistics guy in my type of job would be responsible for just one system. I ended up having three or four at any given time, plus all the related support equipment involved.

For example, when I was working a wheel brake parts problem, I had to learn all about the tires too. It really was fascinating, because the more you learned about the subsystems of this aircraft, the more remarkable it became. I learned that each tire had to be handmade, and it was very critical that they be constructed during periods of very low humidity, so the rubber would mold just right. These were the first tires I had worked with that had no real grooves in the rubber, just slight

dimples. Beale once reported a problem with tires blistering, and we traced it to a high temperature condition in the tire storage facility. They required pretty delicate handling. Operationally they were filled with nitrogen to 400 psi and were good for about fifteen landings. Trying to get all three tires on one main gear, filled to the same pressure at the same time, was fairly tricky for the maintenance people.

We always enjoyed getting on location with the jet since it gave us a much clearer picture of some of the problems the operators were having. The aircraft was like one big puzzle that would only reveal its secrets one at a time.

Take a simple thing like a hydraulic leak, for example. On this plane the hydraulic fluid wasn't reddish in color as it is with most other aircraft. It was basically the same color as the fuel. Here you have a plane which leaks fuel profusely but is not supposed to leak any hydraulic fluid, but we've got to find that hydraulic leak. Not an easy task. So there we were one night in the hangar, going through every inch of those lines trying to locate that leak. It was cold, and our shoes were soaked from sloshing through dripping fuel. And no matter how many lights we'd try to beam into

Aborting a takeoff was risky business.

that black beast, it was extremely difficult to see up into certain critical areas, which of course were precisely where the leaks normally were. Well sometime after midnight we located the problem and we actually let out a cheer. I guess it sounds somewhat ridiculous now, but that's how this plane was. It got right into your very mind and, in time, became a part of you.

Normally, we were called up to Beale only when all hell broke loose. I remember once when one of the jets aborted on takeoff and actually had to take the barrier at the end of the runway. Well the cable that was supposed to help stop the aircraft didn't work just right and came up over the cockpit. Keep in mind, this whole barrier system was something the Air Force really didn't want any part of. They felt it was somewhat useless and too hard to maintain. So up to that point, we never had much interface with the Air Force concerning the barrier system.

Well they couldn't call me fast enough when it didn't work right. From our engineering standpoint, we knew the system was a good one and we considered it a life-sustaining piece of equipment. The SR-71 was not equipped with a tailhook, so this was the best way to stop the jet in an emergency. This barrier system was the only one of its kind and therefore required special maintenance.

Basically what we had done, was to combine the two main Air Force cable systems into one heavy-duty stopping device. If an SR-71 had to stop on the runway due to an aborted takeoff, or make an emergency landing with no brakes for example, then this would stop the plane from plowing into the weeds past the end of the runway. When the system was activated, an inch and a half of steel cable would be thrown up from under the runway and catch the jet between the nose gear and main gear in a matter of milliseconds. As the cable played out from the aircraft's forward motion, the jet would be decelerated, hopefully, by the tension of the cable. We had designed the system so that the cable tension and positioning around the landing gear would help keep the aircraft straight on the runway. The whole thing was an unbelievable contraption consisting of compressed air, electronic sensors, mechanical switches and even some hydraulics.

Keeping the barrier maintained was difficult, and of course it was very rarely actually used. There is a government regulation concerning Air Force barrier systems that requires barrier testing to be done every so often. Well, due to the complicated nature of this system, someone decided that we were just going to forget about that, because we never tested the thing on a regular basis.

When I came up to Beale to inspect the barrier following that one errant engagement, we had to go into the underground pits of the system to inspect the electronic components which activated it. When we removed the metal plates which sat flush with the runway, we found several families of very healthy rats inside.

Needless to say, this did little to bolster the Air Force's confidence in the system. Their basic feeling was that since the system was down for maintenance so often, and SR-71 flying continued anyway, it really wasn't that important. They got away with this type of thinking mostly because

Pilot's view from hangar.

Exiting the shelter with usual crowd of support folks.

178 THE UNTOUCHABLES

the jet was pretty reliable on takeoff and rarely had to make a high speed abort. And with that one exception, on those rare few occasions when the barrier was actually used, it worked as advertised.

The thing that now sticks most in my mind about the barrier is that barrier maintenance crew at Beale. Here we had a system that few people understood, many didn't even want, and some didn't even know existed, and yet by regulation that whole system had to be maintained on a full-time basis. The guys who maintained that barrier were probably the only people on base who knew if they were keeping it in working order or not, and with all that, when I would go up there for my inspections, you have never seen a more committed, enthused group of men. To them, it was their link to that Blackbird, and they took immense pride in being a part of it.

I can remember standing out on the runway late one night, asking them to remove the panels covering the cable — quite a job for that hour. I never heard one voice of complaint, as I suspected I might. Instead there was this eagerness to show me how they had kept the pits cleaner than before and that "their" system would work when called on. I couldn't help but think that these guys were truly among the many unknown soldiers in a program that reached so far beyond the end of that runway at Beale.

— Jaime Gough —
Technician, Configuration Control
8 years

On scene tech rep.

THIS BOOM'S FOR YOU

FROM THE FRONT SEAT . . .

DAY THREE
1435

The debriefing brings good and bad news. The main camera did not produce satisfactory pictures, due to a major component failure. It is a rare type of failure and didn't show up on any of the self-tests. Thankfully, the TEOC cameras worked perfectly and covered the area well.

Walter, being the pro that he is, is immediately concerned with trying to figure out what else he could have done to prevent the failure. There was nothing. The tech reps even say the fact that he was able to get it to work at all in the condition it was in is amazing. I am glad the backup cameras worked, but it is not so easy for Walt. He will stew about it for several hours now, going over every step in the camera checklist to ensure he didn't miss something. Eventually he will be satisfied that there was nothing he could have done to preclude its failure, and lose that very serious look on his face. I am glad I have a guy like Walter in my back seat.

We are told there will be another flight tomorrow. These are exciting and meaningful missions, but our fatigue has drained away some of our enthusiasm over this news.

The commander gives us the names of the men lost in the F-111. The WSO was one of Walt's friends. We didn't know the pilot very well. There is not much we can say. I know Walter has had a rough day.

We pour over mission planning materials which have now become quite familiar. There will be a new twist to tomorrow's route, however. We will fly in a counterclockwise direction through the target area instead of the clockwise routing we have used twice. The mission recorder tapes from our jet show so much threat activity, the planners think it wise to change the direction in hopes of being less predictable to the enemy.

We will fly the backup, as we did on the first day. Bernie and Denny listen intently as Walter describes the DEF panel indications he observed. This will be their first run into the actual target area. As Walt and Denny confer over a pile of checklists, Bernie looks at me and asks if I saw anything of note out the window. I tell him no and he says that's good, because he probably won't look outside at all as he'll be double-checking his cockpit for snakes.

1810

As we drag ourselves to our rooms, we can't help but notice the protesting groups outside of the base, and the combat-like atmosphere that continues to permeate our living quarters.

I eat some peanut butter on crackers and turn on the television. On every channel is continuous coverage of the events of the past couple of days. Of particular interest to me was the cable news channel's taped coverage of Colonel Qaddafi himself, speaking to a large gathering of people, spewing out more anti-American rhetoric. As he was talking, the muffled sound of explosions could be heard in the background. Qaddafi immediately discontinued his grandiose stance as defiant warrior and scrambled quite unceremoniously for cover. The TV news reporter explained the taped scene by saying that instead of another bombing attack, she had good information that the booming sounds heard over Libya were those of an American SR-71 spy plane which had traversed the area earlier that morning. Of course, that accurate statement by the newscaster was then followed by some conjecture on her part about the SR-71's capability to carry bombs.

I shouted to Walt next door that the "Colonel" himself had heard our sonic boom. Walt joined me and we both shared some snacks as we stared with childlike interest at the TV. This is somewhat of a first for us, to actually "hear" ourselves on the news. With some quick channel changing, we are able to see the same report repeated. Watching Qaddafi leap from his podium makes us laugh and I am glad to finally hear some cheer in Walt's voice.

The distinct look of fear I saw on Qaddafi's face when he thought a second strike was in progress told me that, all of his idle threats aside, we had negotiated our message to him loud and clear. His threats to American incursions across his "line of death" seemed, now, hollow and weak. I went to sleep that night gratified that I had boomed his shorts off.

Typical aerial photo by SR-71 camera system,
taken from 78,000 feet.

THE MISSION *181*

From the Rear Seat . . .

Day Three
1345

I knew something was wrong when, after shutdown, I saw the camera tech rep shaking his head in despair. Apparently the OBC had come off its gimball mounting just enough to render the pictures unusable. And as we have become so accustomed to hearing in this program, it was the first time anyone could remember this type of failure occurring. Luckily, the TEOCs did their job perfectly.

After getting out of my suit, I went back over to the hangar and conferred with some of the sensor guys there. We went over everything, and I finally figured out that there wasn't much I could have done about it, but it didn't make me feel that great anyway.

We had a few minutes before the main debriefing, so Brian and I just sat and relaxed in the mission planning room for a few moments. Mostly I felt a sense of exhilaration that we had been "there and back." It is always like that when you fly into the jaws of danger, and return. I always felt that we could take this jet just about anywhere, and we did. After going in to get the battle damage assessment, I felt somehow connected now to my buddies at Lakenheath who had bravely gone in first.

While awaiting word about further missions, I am thinking about calling my wife back home and I hope the few available phones are not tied up later. I learn from the commander that the WSO who was killed in the raid was my friend and fellow flight evaluator from my days at Cannon. It is a sad thing for me. He was one of the really good ones. In this business it seems to always be that way. His wife would never receive the call she so badly wanted to hear.

We find out that another mission will go tomorrow. We'll be the backup to Bernie and Denny. After the threats we encountered today, the mission planners decide to hit the target area from the opposite direction. It seems that the people in Washington, who are daily viewing our results, are happy with the assessment of the bombing raid but want more pictures of Qaddafi's actual headquarters. We learn that the Air Force Chief of Staff is on base and has said he wants to personally congratulate everyone at the DET for a job well done. Our ops officer also tells us that we will all be getting medals for our part in the operation. We never saw the Chief of Staff or any medals. In light of all that we were feeling at that point, both meant very little to us and, in retrospect, not having seen either never diminished one bit our personal satisfaction and recollections of our part in the operation.

1705

I called my wife and learned that my daughter had heard on CNN News that an SR-71 had participated in the raid on Libya. She of course had lots of questions for her mother, who had told her everything would be OK, and luckily for us it was. It was my daughter's birthday.

After mission planning, I return to my room and am visited by some of my F-111 friends from Lakenheath. They talk now, openly, about the fear they had experienced during the mission, and some of the surprises on the raid. Their route had been exhaustingly long, and once in the target area, they had relied heavily on their speed and element of surprise. They speak in hushed tones about their two comrades who were lost on the mission. They are no longer the same people I saw a week ago at the O'club. They are older and wiser now. It is a dangerous business.

Before getting some sleep, I hear Brian shouting to me from his room. As I enter, he is talking to the TV. Fighter pilots do that sometimes. He is murmuring something like, ". . . lady, if we'd been carrying bombs you can bet they would've heard more than sonic booms. . . ." We are both emotionally exhausted and find some comic relief in the irony of watching the results of our flight on the news.

Watching Qaddafi duck his head as he heard our sonic boom echo across the desert made me laugh out loud. It was strange, getting to see it on the TV. For so many missions prior to this one, the "take" area was simply a piece of geography below us, and except for the usual electronic signals in my cockpit, there was no indication of human reactions below. I was quite satisfied with this particular reaction that I now got to see on the news.

With force.

THE CRITICAL FORCE...

I had a rather unique position in the program, as I was a civilian employed by the Air Force Logistics Command, and I got to stay on location with the plane all the time. I didn't know a thing about this plane when I was interviewed for the job in 1965. The person interviewing me said he couldn't tell me where I would be going. I eventually got him to tell me it was in the States, and when he said it would be somewhere in northern California, I was more intrigued. Once I was fully briefed on the program, I was very enthused. I especially liked the idea of working out on my own in the field. I had done that in Arizona with the old Titan missile system and found it suited me much more than any office ever could. Besides, once you fired those missiles off, they were gone; this thing came back.

At Beale, I worked with both the Air Force and civilian side of the house. It was quite a dynamic program to be a part of, since everyone was allowed to perform their job unencumbered by the usual bureaucracy, and our inputs were not only heard, but acted upon quickly. This was due in large part to Paul Mellinger, a real catalyst to the program in its early days and a key man to have on location throughout the years.

It was not unusual for us to issue a priority request for a C-5A transport to haul a certain part for our jet to Kadena, and we'd get it, too. We soon learned who in the chain you could count on to get large amounts of equipment moved great distances on short notice. There was never any question about calling people at home or after hours, whatever that meant, because we didn't really have any hours. It was pretty much an around-the-clock operation for us, since the program was always growing and there were always new problems to work.

Logistically, this whole program could have easily ground to a halt, but every time we came to another hurdle, exceptional efforts by key people would keep the wheels turning. Keep in mind, we needed types of support and parts that no one else in the Air Force was requesting. No other system in the chain was using liquid nitrogen, for example, in the quantities we were; not to mention stuff like Tri-ethyl Borane. And too, our engine was a closed system in that we couldn't order new ones fresh off the assembly line. We were literally inventing procedures for the ordering of parts for this program. One of my fondest recollections of the program to this day is just how much red tape we bypassed and how much more efficient we were because of that.

There were many subcontractors involved with this plane; companies which supplied a variety of specialized and very necessary parts. I got to interface with all of them at one time or another. Most of them had originally made a pact with Kelly Johnson which said in effect, "We will support this system as long as it is in use." To their credit, most of them did just that, even when manufacturing certain parts in small quantities obviously wasn't very profitable from a business standpoint.

As time went by, the hardest thing was dealing with all the unsolicited "help" the Air Force wanted to give us. We had been getting it done on our own just fine. There were some terrific military people there at Beale, but eventually they had to move those people out, and we were faced with the basic military fact of life, that people continue to move and are replaced with new ones possessing no corporate knowledge.

Needless to say, I was devastated when they announced the cancellation of the program. It was especially hard to understand, considering that we were in the best shape ever, logistically, in 1990. I don't think I have quite accepted it to this day. I never go back out to the base; I really don't care to see the plane sitting on a pedestal.

Most of us were already beyond the point where we could retire, so it really was more of a personal involvement which kept us on with the plane. I give a lot of credit to those initial Air Force maintenance guys who really did the hands-on work. They were highly committed to this aircraft.

My days with the SR-71 program were truly the culmination of my entire career. I always felt like I didn't work for just the civilian contractors or the Air Force Logistics Command, but rather for the entire country.

— *Lee Olson* —
Logistics Field Rep
25 years

THE CRITICAL FORCE...

I first came to Lockheed in 1956 under their Apprenticeship Program. This was a four-year school which, after graduation, qualified me to work in just about any department at the plant. I learned a lot about many different areas, from hydraulics, to electrics, to fuel systems, and much more. As an apprentice, I was authorized access to most of the plant and one day walked through a finishing area where an A-11 sat. I was awestruck. I had never seen anything like that before.

My first job after training was working with the U-2 program. Once I got out into the field, I really enjoyed it and learned to work well with the Air Force maintenance system. My official title was travelling field service representative. On paper, my area of expertise was fuel systems and aircraft structures, but I ended up doing so many different things most of the time.

I came to the SR-71 program in 1965 and stayed for just about its entire operational life. I enjoyed most the different people I had the chance to work with. We were always faced with trying to solve one problem or another. We were much like detectives in that sense, sometimes feeling a little like we stood between the company and the Air Force, when we eventually would deliver our findings. I felt there was a wonderful trust between the Air Force maintenance folks and the tech reps. The military guys understood that we weren't as interested in assigning blame as we were in researching a problem to preclude its reoccurrence. We all seemed to have the best interest of that airplane as our foremost concern.

The difficult part of my job was the time I had to spend away from home. The TDYs to the DETs for me were averaging about three months at a time. I was on the first trip to help set up DET 1 in Okinawa and also made the first trip to put DET 4 together. Altogether, I made eleven trips to England and twenty-nine trips to Okinawa.

I was also sent on quite a few trips to help recover jets which had made emergency landings somewhere. Because of my familiarity with SAC procedures from my days on the U-2, I could interface with the Air Force quite effectively concerning a variety of procedures, and that was probably why I was sent on more trips than anyone else. Paul Mellinger also always sent me on the really difficult trips.

In 1967 I had the dubious distinction of being sent out on the first accident investigation team when we lost our first SR-71 from Beale. The crew on that one had just completed night aerial refueling, and after coming off the boom, the pilot experienced severe spatial disorientation. About the time he realized it, the gauges were spinning wildly, but he still felt like he was straight and level. Both he and his RSO ejected safely, but by the narrowest of margins. We calculated that they exited the jet approximately eight seconds prior to the aircraft impacting the ground.

The plane was going just about straight down when it hit the ground, so there weren't very many large pieces for us to inspect, but I do remember a couple of things about that trip that make me laugh when I think about them today.

The jet went down in the mountains of northern New Mexico, and the first person there to greet us was the rancher who owned the land. He said, "I hope you guys aren't planning on making this a habit." Apparently, some months prior, the test facility SR-71 that went down also happened to crash on this same man's land, not far from where we were sifting through the latest wreckage.

The other thing I remember most about that trip was our having to literally shoot those J-58s to "kill" them. Of course, those engines were smashed beyond useful repair, but they were tough enough to retain in one piece the small attached tanks of TEB. We knew we had to get rid of that TEB before it created a hazard to someone. One of the security guards assigned to the site had driven his camper out to the location and had with him a high powered rifle. Well, we just stood back and shot those tanks. We knew the liquid nitrogen keeping them pressurized would force the TEB out and it would burn out on contact with the air; and that's just what happened. I'm not sure we'd do it exactly like that today, but it worked. The rancher remarked that it was a bit like shooting a horse with broken legs, and I guess in a way it was.

Altogether, it was quite a program and quite a unique airplane. It bothers me to see the weathering and debris on the ones on static display today. She was much too proud a jet for that.

— *Jack Boen* —
Tech Rep
24 years

Display jet at Beale AFB.

CHAPTER 5

The Final Run

From the Front Seat...

17 April, 1986
Day Four
0520

Suiting up for the third mission was more of an arduous task than the normal meticulous dressing routine. I felt like I literally poured myself into that heavy garment, and the helmet had somehow, surely gained weight. Once the suit-up procedure was complete, I sat, motionless, in a large reclining chair, and couldn't remember being out of that space suit in the past forty-eight hours. There is a part of me that is glad we are flying the shorter mission today. The physical demands of the past few days have begun to catch up with us.

Brian tries to connect spurs to seat reels.

As we approach 960, she looks lean and hungry to fly, as always. But I know she is hurting, as there are numerous minor discrepancies the maintenance crews have not had time to work. I know she gave everything inside of her yesterday, and I hope she is feeling well today.

The people in the hangar look tired and unshaven. They have been working around the clock to keep both jets flying each day. Everyone is giving a maximum effort and no one is complaining.

I see Doc and a couple of the guys whose familiar faces have become a part of our TDY life. As always, they greet us with enthusiasm and it makes us feel better about everything they've done, or haven't had time to do, with the plane. These are truly dedicated people. As we take that long climb up the ladder, I hope my fatigue does not show.

As I stand on the cockpit floor, I try to engage my boot spurs into the small sockets below the seat. It is done by feeling the metal joint with your heel and stepping down into the slot. Though it is a routine accomplished on every flight, I take an inordinate number of stabs with my heel spurs until they are both connected. In case of an ejection, this will ensure my legs are reeled back into the seat instead of flailing painfully in the slipstream.

After the primary jet launches into the dark sky, we sit in our cockpits waiting for our appointed takeoff time. The usual swirl of gray-tipped clouds has enveloped the field, and in open

Resting comfortably in PSD van.

cockpits, outside the hangar once again, we just sit and wait. I notice the civilians kneeling and standing around the jet, just waiting too, everyone so motionless.

I stare at the blank canvas of gray sky above us. I know it will be clear above. My body feels like it has taken on the shape of the aircraft's seat. I notice several news teams huddled outside the perimeter fence, filming the takeoffs of a couple KC-135s.

Before putting the face plate down, I could feel the cool morning air blow across my face, heavy with the feel of impending rain. I couldn't help but think about the many pilots who sat like this many times, not too long ago, at numerous fields across this same English countryside. Sitting in their Hurricanes and Spitfires, the aircrews, too, must have felt similar exhaustion as they waited for word to launch on yet another mission. Hurling their simple machines into the same kind of ominous weather so common in East Anglia, they too, had propelled themselves into the unknown of the day's mission. The technology had changed, but I felt certain that many similar emotions occupied the cockpits of both eras.

The brutish howl of the start cart revving up interrupted my thoughts and I locked the faceplate down.

FROM THE REAR SEAT . . .

DAY FOUR
0545

No matter how many hours of sleep I was able to get, the third day was a bit of a push for my body. I had a feeling that this could be the last sortie of the operation, and I was hoping it was. In three days, we never really had time to unwind, to "come down" from the missions. The amount of adrenaline which flowed each time we flew this plane would always catch up with us eventually.

I am most impressed with the maintenance crews who tirelessly continue to produce two jets each day for us. This is not an easy jet to maintain. Their words of encouragement to us each day we see them in the hangar have become more important to our morale than any official Air Force pat on the back. I have rarely experienced a more intense example of team effort than I have during this operation.

I feel good about the jet because I know from experience that the more you fly them, the better they perform.

While reclining comfortably in the chair in the PSD van, I momentarily fall asleep. With my green visor down, no one really notices and I sleep soundly for three or four minutes. Somehow, my brain sends emergency signals to my tired body, and I awake just prior to when we have to climb into the plane. I eat a little tube pudding in the van, and it helps.

Taxiing out. Spare tankers in background.

DET 4 departure.

From the Front Seat . . .

Day Four
0652

We plow subsonically through sheets of rain and finally get to our first set of tankers. The entire process has become routine now. Same tanker, same bad autopilot, a little turbulence, and the same roller coaster chase. I fall off the boom repeatedly; partly because the tanker is unstable, partly because I am exhausted. Walter patiently sits through the process as he knows saying too much helps little. Somehow, I manage to get the required gas. Walt calmly tells me that we can get off the boom a few seconds early and still have enough gas and that is nice to hear.

The jet moans and groans all the way to Spain, a carbon copy of yesterday's antics. She needs a little down time, but we all do and she gets little sympathy, but I continue to stroke her gently with the controls as it is the best way to treat this lady. The outside air temps are better today and our fuel looks good descending into our second tanker.

The tanker guys are quite talkative on the boom, and since they have been rotating crews, they sound more rested than I think anyone has a right to at this point in the week. Someone is taking a picture of us from the boom pod. I try to smile.

Somewhere on the refueling track, as we are snuggled neatly under a KC-10 at 29,500 feet, the fatigue monster creeps into my cockpit and tightly wraps himself around me. The simplest movement of my hand on the stick is an effort. I feel my right forearm cramp up, and for a few moments, fly with my left hand. I know it is a combination of fatigue and dehydration. I have lost six pounds in the past three days and had very little sleep. I move around in the suit as best I can

Waiting for engine start.

to relieve the cramping in my arm and shoulder. No matter what I do now, my arm quivers with muscle tremors and there is a constant pain in my shoulder.

My water bottle is already empty as I have sucked it dry less than two hours into the flight. I feel a certain sense of relief when Walt receives the radio transmission that tells us the primary jet is a go, and we can turn back toward England.

We turn toward home and hope that Bernie and Denny are experiencing an easy run. My map display screen fails after a few moments at Mach 3.1. I really don't care as I rely more on my knee map anyway. The little map that I draw for myself gives me a bigger picture of the whole route and helps keep me oriented better than watching small segments of the projected map display move incrementally along the route.

I am exhausted, and the jet is kind as she cruises home at 2,100 miles per hour with no new problems. Ol' 960 has been a good jet throughout it all. Speeding smoothly toward Land's End, she shows little of the fatigue which engulfs me and seems to be proudly telling us that she can continue to fly for however many days we elect to continue this operation.

Walt and I talk little on the return as we are both somewhat drained from the past few days' events. We are greeted with uncharacteristic sunny weather back at Mildenhall. Under these conditions, there are few more beautiful countrysides than the rolling green hills of East Anglia, and the scene gives me a shot of renewed energy.

Walt compliments my landing and I feel good about that. As I pull a yellow lever in front of me, the large drag chute deploys and blossoms behind the jet, pushing us firmly into our restraining straps. As the jet slows, I see the familiar mobile car passing us on the runway, and I shove the drag chute handle forcefully inward, jettisoning the large chute behind us.

As I clear the runway and head back toward the SR-71 shelters, I see that familiar group of men who have made all our flights possible. To the uninformed visitor, they might appear a motley crew of spectators dressed in a variety of coveralls, jeans, and parkas. But to us they are our vital link to all that makes this jet powerful and magical. Weary with fatigue, after three straight days of servicing, loading, downloading, fixing, inspecting, cleaning, and double launching, they stand now, weary but proud, along the sides of the hangar. As always, there are the enthusiastic thumbs

Crew chief prepares to recover jet.

up and that look on their faces that always conveyed "We're glad you're back." As I slowly eased the stiletto face of the Sled across the fuel soaked hangar floor, I felt as if Walt and I were being received by the highest ranking VIPs in the program.

After pulling the throttles back from the idle position, I disconnect my gloves from the suit, and after nearly four hours of breathing 100 percent oxygen inhale large breaths of fresh air. My hands are sweaty, and my legs feel slightly weak as I stand up in the cockpit. When the helmet is removed, it is a wonderful feeling, and scratching my head takes on new pleasure.

At the bottom of the ladder, as always, are the commander and ops officer waiting to hear about any unusual occurrences we might have experienced. There are also some systems specialists eager to hear if any of their components went awry. Doc waits patiently, as he always does, to hear

One exhausted pilot.

if there is anything we need to tell him about his engines. I am happy to report to him that, once again, they are error-free, and the look of satisfaction on his face is one of my most lasting memories of the entire operation.

Everyone looks exhasted. As I crawl into the PSD van, I know that even as we drive toward the squadron building, our guys are readying the jet in the event we are tasked to fly tomorrow.

From the Rear Seat . . .

Day Four
0820

This is a first for us — three consecutive days of flying. My space suit now feels slightly heavier than normal. I am glad I am not the pilot today. The refuelings have been very physical and I know Brian has to be tired. I try to occupy myself with the ANS and some sensor checks while Brian goes through that first AR. This helps me feel less apprehensive about the heavy breathing I hear coming from the front seat.

The jets are holding up well and I feel confident that Bernie and Denny will complete their run. When I finally receive the message to return to base, I am a little relieved. I am exhausted so I know that Brian must be even more so at this point in the week. I have kidded Brian in the past about the amount of time he spends working out at the gym, but I am actually glad that I fly with a guy that does.

The trip home is routine, though routine is a somewhat relative term when referring to flight at three times the speed of sound. Things are either very smooth, or I am engaged in performing the world's fastest checklist scanning. The inlet still buzzes and the resulting vibration is disconcerting, but we are almost immune to it now. The plane is otherwise solid and once again delivers us safely from another mission.

Kelly Johnson was instrumental in the design of over 40 leading aircraft, and since 1937, had won nearly every award possible in aviation: The Wright Brothers Medal, Medal of Freedom, Spirit of St. Louis Medal, the only man ever to be awarded the Collier Trophy twice, and the National Medal of Science, just to name a few. So revered was the genius of this man, that in 1973, a new aviation award was created in his honor: the Clarence L. Johnson Award. The first recipient of the new award? Clarence L. Johnson.

THE CRITICAL FORCE...

In the early '60s I was working as a field rep for Lockheed in the F-104 program in Japan. I wanted another job with the company and had the choice of going to missiles at Vandenburg or to a "mystery" program. I chose the latter and was introduced to the Blackbird family in 1964.

I was very impressed with it initially, even though I had not been briefed on any of its performance. When I got to see it fly at the ranch, as an engineer the two things that impressed me the most about the plane had nothing to do with its enormous speed. I was extremely impressed that this aircraft could survive, and stay in the temperature environment that it did. I was also surprised to see just how fragile it was in certain sections concerning the G limits. If you were at high speed and low altitude, it really wasn't too difficult to over-G the plane with an aggressive pull on the stick and snap the spine of the fuselage.

Originally, my job was to prepare and give classified briefings on the performance and parameters of the aircraft. Only people who had a real "need to know" got the briefing, and in those days, this included very few Air Force people. Besides at the plant, I would also give these briefings at various other locations, and this meant I had to transport the classified 35mm slides for my briefing. Those slides were never far from my reach. I had a special seam sewn into my jacket, and the small box of slides rested there. That jacket went everywhere I did. When I went to sleep on a trip, those slides were always within arms reach. That's how important the security was then.

As Lockheed employees, we were searched every day upon entering and exiting the plant. If you lost your badge, you were out of a job. The security of the program simply was not to be compromised. Kelly was very good about including only the minimum number of people with information about any given area. If someone really didn't need to know something, they didn't. Kelly also gave those with clearance full opportunity to use all of their skills. This resulted in people producing an incredible degree of excellence in their work.

There was a terrific amount of trust in the early days between the suppliers and the operators. There were basically three categories to the procurement of a new weapons system into the inventory. In simple terms, CAT I involved the guys who built it (contractor), CAT II included the guys who ensured it was acceptable (test force), and CAT III concerned the people who were going to use the product (operators). I would say without any doubt that the Air Force has rarely, if ever, experienced such a smooth and efficient transition with any new system from procurement, to testing, to operational status, as they experienced with the SR-71 program. This was due in large part to the way we were allowed to do our jobs.

Many times, the three categories overlapped somewhat. Here was a plane that had been operated by the CIA and was now going to be incorporated into the Strategic Air Command. Can you imagine trying to do something like that today? The system actually became operational with the Air Force much quicker than anyone would have predicted at the time, and all with a minimum of problems. This required extremely efficient liaison between the people working the different categories.

As I continued to give the classified briefings, I was learning much about the inner workings of the plane, as I would hear the experts at these meetings discuss many questions and solutions to numerous problems.

Although the aircraft was built with some stealth design features, that aspect of it was never fully developed. The general consensus, especially in the early days, was that the aircrafts' speed and altitude were sufficient to protect the crew from hostile threats.

I've heard many people remark about how rough the surface of the plane is, especially at some of the joints. Most people assume that the great speed would dictate extremely smooth aircraft surfaces. Actually, at the speeds this airplane flew, the actual airflow is not right on the skin of the plane but slightly above the surface of the plane in a sort of boundary layer of air, following smoothly the aerodynamic pressure patterns which Kelly had predicted so accurately. The expansion joints and other seams on the plane, though giving a slightly less than flush surface, actually had no detrimental affect on the performance.

We also started out with rudders made out of a material closely resembling cardboard, but eventually went to the all metal construction. Although it has been declassified now that the recommended top speed of the jet was listed at just over Mach 3 in the pilots' manual, that really wasn't the limit of the aircraft. Its limit was really measured in degrees of temperature, primarily concerning the inlet. With certain atmospheric conditions, that plane would go about as fast as it wanted to. To this day, this airplane is in a category by itself for its ability to live in that temperature environment.

There is no doubt that Kelly was an aerodynamic genius, but he was also a great leader of people. He knew how to manage the brilliant engineers that worked with him, and that was rare. By their very creative nature, engineers are difficult people to manage. They start with a blank piece of paper and then develop drawings and ideas that reflect their very best talents. They really don't like having someone come back later and explain to them why their idea isn't so good after all, especially when they are some of the best qualified people in their field. Well, Kelly was a genius at handling his group of engineers and was able to show them where something wasn't quite right, and he could actually make them like it. He could work them sixteen hours a day sometimes. His personal drive to make the best and do it first, was very motivating to all those around him. He could be tough, too, and was a bit of a taskmaster at times, not tolerating oversights very well, but he was such a great patriot that everyone who worked with him felt proud to be on his winning team.

The marriage of the engines, the inlet, and the airframe was quite an engineering feat. The inlet system was the most critical. It was what would enable this plane to cruise at higher speeds than any other air-breathing jet in the world. The air entering the inlet had to be decelerated greatly before the engine could digest it, and that spike had to precisely control the shock wave. These were not easy engineering hurdles, but somehow each problem was solved, usually after much blood, sweat, and tears.

As the plane continued with more test flights, another new finding was that as the fuel burned down in the fuselage, the top of the plane got hotter than the bottom which was kept cooler with the remaining fuel. This caused a differential in the minute warping of the airframe at high speeds and needed to be addressed when adding the various sensors which the SR-71 would carry. All of these types of engineering problems had to be conquered before the jet could fully perform its mission. Keep in mind, too, that all of this testing, developing, and problem solving occurred in a relatively short time span, so there were many long hours by many talented people.

Sometimes, the engineers at the plant were a little frustrated about not being able to tell people about their work, and getting very little feedback from the user. I set up some tours for these folks so that they could get out in the field for at least a day to see the plane fly and talk to the maintainers. They loved going to Beale. Those visits really gave them a perspective unlike anything they could get at the plant.

When the first SR-71 squadron was formed at Beale, the commander there indicated that he would like to have a key Lockheed representative on base at all times, and thus the job was created which I would fill for the following twenty years.

Every time the plane flew at Beale, as the senior Lockheed field rep, I submitted a report to Burbank detailing any problems or fixes necessary. Occasionally I would go TDY to one of the DETs, and I liked this because it got me closer to the plane and out of the office more than I could ever do at Beale where I was constantly working some issue. We all used to enjoy watching the launches, and we were always there in the hangar when a plane came back from a flight. We were anxious to hear about any problems the crew might have experienced. I always had the highest respect for the men who flew this plane.

I guess my favorite memories of the program were during the middle years of the SR-71's operational life. We were flying four to seven flights a day and had a nearly perfect working relationship with the military at all levels. We were able to talk on a professional level with everyone regardless of their rank. People were willing to listen to those with the most system knowledge, which usually meant listening to the civilians' inputs, and in turn, it made the military more efficient. We were very responsive to the military's operational needs and there was mutual respect all around. Everyone was on the same page with the same goal of keeping the mission going, safely and successfully. Everyone was fully confident of everyone else — a unique situation to be sure. When questions arose, everyone knew that the answer given was not just something to appease the "old man"; it was the correct data. It was sad to see the Air Force change in that way over the years, but we never did, and I suppose that's why eventually we made some people feel uncomfortable.

There were times, just by the nature of my job and the type of decisions that I could make, that I know I made some military folks feel a little insecure.

When we moved into the era of DAFICS, we felt like we were starting all over to some degree. Clearly, this was a much needed and improved concept, but mating '80s technology with a '60s airframe proved to be a challenge. The aircraft was basically designed for analog systems, and now we were installing a completely digital system to monitor flight controls and inlets.

One of the major problems was, the digital systems count pulses. The computer doesn't really care if these pulses are true signals or noise pulses; it counts them all. Well, this aircraft was full of noise impulses, and these were being added to the signal impulses, causing havoc within the system. The aircraft had to be "cleaned" electronically. This was a bit of a nightmare for the folks doing the rewiring. Once the bugs were worked out, the plane was flying more efficiently than ever and this resulted in some fuel savings, something the pilots were happy to see. Again, the beauty of the program was that the entire transition to DAFICS was accomplished with no negative impact on mission effectiveness.

Gradually the Air Force had lowered the skill levels required to be assigned to SR-71 maintenance. This resulted in Beale getting some very inexperienced maintenance folks. This made our job as tech reps even more critical, but sadly, the Air Force felt that by this time we had pretty much outlived our usefulness to the program and there was a definite change in attitude. I can remember quite clearly one moment in time which for me crystallized the whole change in attitude by the military toward the tech reps. We had always been more like a big family at Beale, the tech reps and the military doing a lot of things together, both professionally and socially. One night I was in the officer's club, as I had been many times before, and one of the colonels saw me, and with a look of surprise, said, "What are you doing here?" He didn't say it in a rude way, just in a surprised way like, "Why would you even want to be here?" The changing of the guard was never more apparent to me. I really didn't mind. I knew that all things eventually change. My concerns were more with the jet, and I had to recognize that for some people associated with the program, their main concern was their promotion.

In the last years of the program, SAC mistakenly felt that the SR-71 should be treated like any other plane in its arsenal. I spent most of my time trying to train and retrain people who were new to the program. Due to the inexperience by then, at many levels of the program, there were certain instances where we came extremely close to experiencing a catastrophe.

I don't think SAC was ever totally comfortable with having a reconnaissance system in its arsenal. Most of the higher brass up at SAC headquarters were bomber guys, and this airplane was definitely no bomber.

Some folks I worked with were on this program for basically their whole career. Nothing followed it. The company was ordered to destroy all the tooling and presses, so no follow-on models of the plane could be developed.

Looking back now, I feel like we were all a part of a rare page in aviation history. I loved working around so much enthusiasm and especially being allowed to use my talents and imagination while in the field. When good, qualified people were allowed to exercise their talents and make decisions, the program worked beautifully. You'd think that maybe someone would have learned a lesson somewhere in all of this, but it seems they haven't.

The most satisfying thing to me was that we never lost a crew member in all those operational sorties flown over all those years. I'm very proud of that. I guess the challenge of being a part of it all kept me young. I am 73 years old now and can't remember that many years going by.

During the Gulf War, we all realized what a mistake the cancellation of this program was. I kept waiting to see the replacement for the SR-71. I'm still waiting.

— *Paul Mellinger* —
Senior Field Rep
23 years

Margaret Martin and Paul Mellinger. Two patriots.

MISSION ACCOMPLISHED

From the Front Seat . . .

DAY FOUR
1500

The debriefing goes well. The plan to fly the primary jet from west to east over the target area seems to have worked, as the other crew reports little missile activity. Finally we are told that no further sorties are required and we are to take some days off. We are relieved, and Walt and I discuss whether we should go eat or sleep first.

Before we leave the squadron, one of the intelligence officers calls Walt and me aside to let us know that we had two confirmed missile launches on us yesterday and one probable. One missile missed and splashed down harmlessly in the Gulf of Sidra. The other misguided and fell back to earth; the wreckage was now being displayed by Qaddafi as "evidence" of the U.S. bombing of civilian areas. We do not ask how this information is obtained, as we know he wouldn't tell us anyway, but we appreciate knowing such details.

Our "normal" mission toward the Barents Sea will go as scheduled next week, and it is now time to get some rest — so naturally, Walt and I head over to the officer's club at Lakenheath to talk with the "Libyan raiders."

Fighter guys are a different kind of people. They hate war, yet will be the first to volunteer to help end one. The business of putting bombs on target is deadly serious, yet there is never a lack of humor in any good fighter squadron. The Lakenheath boys are eager to show Walt and me a new patch design they have developed. In bold print are the letters *L-I-B-Y-A*, which spelled out reads, *Lakenheath Is Bombing Your Ass.* I hope there will always be fighter pilots.

I return to my room and sleep. I do not move for the next thirteen hours.

FROM THE REAR SEAT . . .

DAY FOUR
1500

The operation has finally come to a close. In the past three days we have been in the jet nearly sixteen hours. We've reached our limit of endurance. As we sit around the big table in the main briefing room, we all feel a sense of pride and accomplishment. Every person in the DET has earned the next few days off.

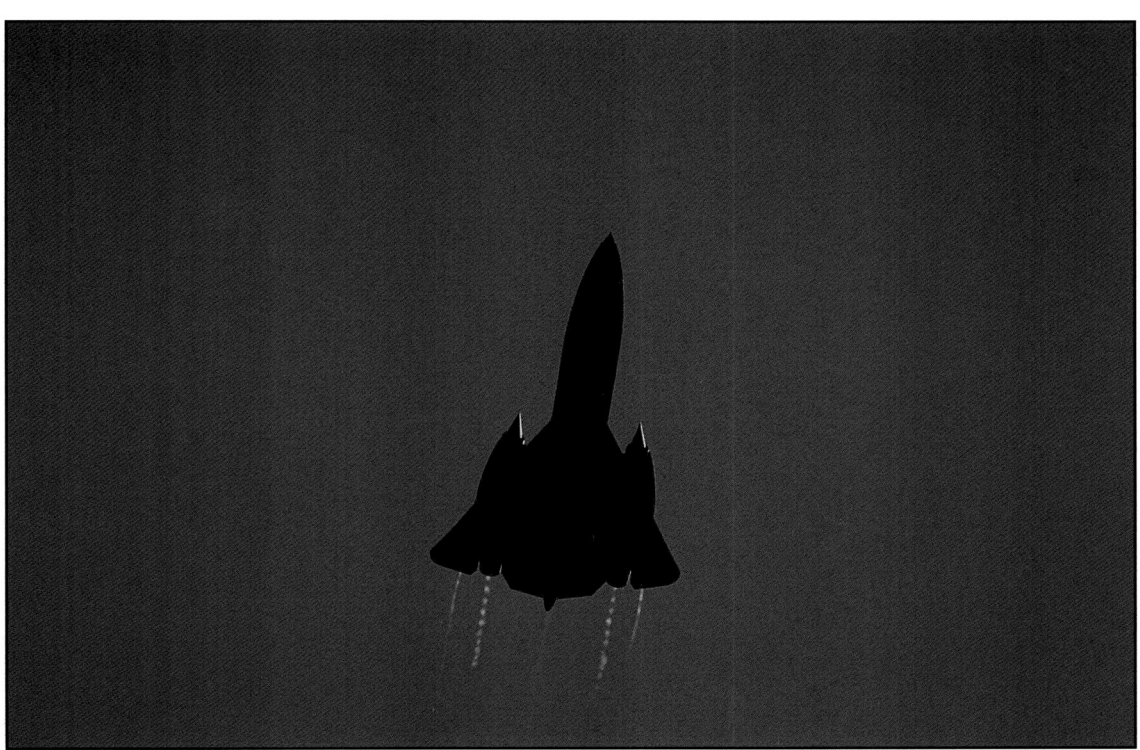

We are briefed that our routine reconnaissance mission next week will go as scheduled. It's one of those 6.3 hour treks up to the north, and though the thought of that right now is out of the question, I know that in a few days we'll both be fresh and eager to fly it.

For now, my immediate thought is to eat. Brian and I overdose on several really bland hamburgers at the base snack bar. I remember they tasted a lot like filet steak at the time.

We need to sleep but are too exhausted to do so, as we need to "come down" from not one, but three, sorties. Our ears will be popping all night from the 100 percent oxygen we've been breathing so much of. So we sit and eat and talk.

The operation as a whole was a magnificent coordination of air power. The base was in total turmoil all week, with protest groups, news people, and tanker crews sleeping in the hallways of overcrowded billets. The reports on British TV have been very interesting to watch, especially for those of us directly involved. I have been extremely impressed with our maintenance people and their ability to meet all necessary requirements.

Later, Brian and I meet with some of my buddies at the Lakenheath Club. They are in a mixed emotional state, between relief that it is over and grief for their loss. They tell me there will be a memorial service at the base chapel next week, for their fallen comrades. Since it coincides with the day of our SR-71 sortie, I request our commander approve our doing a flyover of the chapel upon our return from our mission. Approval is granted.

NOTE: On the final mission of their TDY to DET 4, Majors Watson and Shul, given a choice of two different SR-71s, elected to fly number 960. The aircraft flew flawlessly for over six hours, completing a full afterburner pass over RAF Lakenheath Chapel prior to landing.

THE CRITICAL FORCE...

. . . We were like family working with that program. Even though the plane was not in our building, it was in our hearts. There was one goal — the mission — and there wasn't much that happened to that plane that we didn't hear about.

. . . I was so proud to be a part of this program. I inventoried parts for the engine section. We used to call it, "the thrust you could trust" while you were up there. It seems as if Congress were the only ones who could shoot it down.

. . . The sound of that airplane during a launch could really touch your soul.

. . . In most offices, idle talk centered around kids, vacations, etc., but here we always wanted to talk about the plane, the latest mission, or some new piece of memorabilia being designed for the Blackbird, and we developed quite a few on our own. For us, it wasn't just the most important aircraft in the inventory, it was the only aircraft. No other jet could touch it.

. . . I worked with this program for over twenty years and one of my proudest recollections was being allowed to sit in the simulator. I made sure to sit in the front seat.

. . . We all had that special love for the plane. It was ours, and everybody knew it was the best in the world. I remember quite vividly the first time I saw her in the hangar at Palmdale. She seemed so regal, being attended to by all the technicians. For me, she was always a living, breathing entity.

. . . Seeing it in a museum is like seeing someone you love being sent to prison with no parole. When I saw her like that I just wanted to let her out and see her fly. She wants to breath again.

. . . I finally saw one in a museum, and I think that was my first close-up experience of death. Nothing about her was alive . . . .

Jerry Wilson – Item Manager

Shirley Strong – Item Manager

Carol Wilson – Item Manager

Boe Kopfle – Secretary, Engineering Division

Margurette Hale – Budget Analyst

66 years combined

Epilogue

Before ever starting to write this book, I experienced my first viewing of an SR-71 on display in a museum. I was giving a slide presentation on the plane for the Castle Air Museum in Atwater, California. The museum airplanes there all sit outdoors, and as I drove up I couldn't help but notice the long black jet resting near a tree, apart from the main group of planes.

I arrived in the early morning before the museum was open, so while waiting, I took a quiet walk around grounds. The SR-71 looked out of place amongst the mere mortal airplanes on display, yet I was reluctant to go over to it. I had seen the SR in better times and wasn't sure I would enjoy a closer look at this faded shell of a champion.

With the subtle lure of the fine lady she was, I eventually found myself standing in front of the black jet. With faded decals, she seemed less impressive than I had remembered these jets. She was dusty. I had never seen one dusty. The tires sat slightly deflated, unlike the rock hard rubber I had known. Disinterested, a blue jay perched impudently on the left spike. It saddened me to see the plane so lifeless. There was no seepage of fuel across the lines on her belly, dripping to familiar puddles below. The rotating beacons sat dulled, extinguished now from bursts of red light, and no familiar vibrations emanated from the now hollow spaces where J-58s once nestled mightily. I turned and walked to the museum building, wondering what stories that jet could tell if it could speak.

I can't recall ever giving a better slide presentation on the SR-71. Things I had never remembered to say in previous talks came from my mouth as if memorized, and I talked about the plane with the kind of energy and zeal I had when still flying it. I could feel the presence of the jet in every picture the audience viewed, and when I was through, I felt emotionally exhausted as if I had gone through an actual mission. I felt as if the jet outside was talking to me, just as one had done so eloquently in flight, six years earlier. It was a strange sensation and I attributed it, perhaps, to my seeing, for the first time, an SR-71 displayed as a museum piece. I felt good, though, in sharing my thoughts and slides of the jet with groups like this, and I knew this was the best thing I could be doing to keep her alive. Somehow, it didn't seem so sad that a Blackbird now sat in the museum. Many people would come to see her and maybe appreciate in some way the greatness she represented.

As I was leaving the museum, instead of reluctance, I now felt compelled to once more visit the SR-71 outside. I had accepted her new status and knew it was time to bid farewell to times past. As I stood alone with the Sled at the end of the day, I could easily see past the scars of her weathering and she looked beautiful, as I realized she would always be to those who had been a part of her life in some way. To us, she would never be just a museum piece but a living reminder of all the excellence achieved and sustained for so many years.

As I was about to depart, one of the museum curators joined me at the jet and was anxious to share with me some interesting facts about "our newest addition."

I learned that, apart from the A-11 models, this was the first actual SR-71 to be delivered to a museum. It was also the plane which flew the final operational sortie at Beale during the formal SR-71 retirement ceremony in January of 1990. I was glad to hear that the plane had an impressive combat record too, as the curator informed me that this particular jet had flown more sorties over North Vietnam than any other Blackbird during the war. Its final flight occurred on 27 February, 1990, when it was flown to Castle Air Force Base for delivery to the museum. The jet retired with 2669.6 flight hours on its mighty frame but, more importantly, when finished, she had flown more operational sorties in the service of her country, than any other SR-71.

I was impressed. The jet was *nine six zero*.

The official retirement ceremony, 26 January, 1990.

" *There are people in this country, who work hard every day,*
Not for fame or fortune do they strive,
But the fruits of their labor, are worth more than their pay,
And it's time a few of them were recognized . . . ."

FROM THE SONG "40 HOUR WEEK" BY ALABAMA

210 THE UNTOUCHABLES

THE FORCE

that contributed to the making of this book....

BOB ANTILLA – David Clark Company, Technical Representative, Space Suit

JACK BOEN – Lockheed, Technical Representative, SR-71

JIM COOK – Lockheed, Technical Representative, SR-71

BARRY DeVRIES – Air Force Logistics Command, J-58 Engine Technician

JAIME GOUGH – Air Force Logistics Command, Technician, Configuration Control

MARGURETTE HALE – Air Force Logistics Command, Budget Analyst

MSGT MIKE HULL – United States Air Force, Electronic Counter-Measures

JERRY KEEVER – Air Force Logistics Command, Inventory Control

SAM KELDER – Lockheed, Structural Design Engineer

BOE KOPFLE – Air Force Logistics Command, Secretary, Engineering Division

JACK LEVINE – Northrop, Technical Representative, Astro-Inertial Navigation System

MARGARET MARTIN – Lockheed, Administrative Engineering Assistant

PAUL MELLINGER – Lockheed, Senior Field Representative

LEE OLSON – Air Force Logistics Command, Field Representative

R. 'DOC' STRANGE – Aviall, Technical Representative, J-58

SHIRLEY STRONG – Air Force Logistics Command, Item Manager

NORM SWANSTROM – Northrop, Technical Representative, Defensive Systems

BILL UMBLE – Air Force Logistics Command, J-58 Engine Program Manager

CHUCK WIETHOFF – Honeywell, Technical Representative, Flight Control System

LEW WILLIAMS – Lockheed, Technical Representative, SR-71

CAROL WILSON – Air Force Logistics Command, Item Manager

JERRY WILSON – Air Force Logistics Command, Section Chief Item Management

Glossary of Terms

accel – Acceleration. The term used to describe the climb from subsonic to supersonic speeds.

afterburner – An extended section built onto the engine casing in most military high performance jets for the purpose of increased thrust. Initiated by pushing throttles past the normal full power position, resulting in an extended flame from tail exhaust section. Also called *'burner.*

ANS – Astro-Inertial Navigation System. The sophisticated navigation system on the SR-71, capable of tracking stars in broad daylight. The jet never left the runway without one.

AR – Aerial Refueling.

auto-nav – Automatic Navigation mode of flying the SR-71, when the autopilot and ANS were linked to steer the aircraft automatically on a prescribed course.

boomer – Nickname for the refueling boom operator in a tanker aircraft.

BUFF – Big Ugly Fat (ahem) Fellow. Nickname for the B-52, used mostly by people not in B-52 squadrons, but not always.

comm-out – Communications out, when operations are conducted without the use of radio transmissions.

DAFICS – Digital Automatic Flight Inlet Control System. Installed into the SR-71 in 1983 for improved management of inlet system. Utilized a triple computer system.

DEF – Defensive systems on the SR-71, consisting of an array of electronic sensors and countermeasures housed in large pods uploaded into the aircraft. Panel of readouts in rear cockpit.

DET 1 – Detachment one, Kadena Air Base in Okinawa. First location SR-71 was based out of the United States.

DET 4 – Detachment four, RAF Mildenhall, England. Set up for SR-71 operations in 1983.

DET 6 – Detachment six, located at Norton Air Force Base, California. Provided critical logistical support for SR-71s worldwide.

EGT – Exhaust Gas Temperature, one of many items the pilot continually checks to ensure engines are not overheating.

ELINT – Electronic Intelligence. Not all sensors on the SR-71 were of the picture-taking variety.

FOD – Foreign Object Damage. Anything besides air that is consumed by a jet engine. Typical FOD consists of rocks, runway debris, birds, tools, hats, etc.

inlet doors – Moveable vent openings on the forward part of the SR-71 inlets. By adjusting the door's position to control the amount of air vented around an inlet, the pilot attempted to maintain an optimum pressure in the inlet. Cockpit indications told the pilot door position, inlet temperature, inlet pressure and spike position. Mismanagement of the doors was a quick ticket to an unstart.

INS – Inertial Navigation System. Typical nav system found on modern jets, using geographical coordinates and computer technology to fix accurate position.

Intell – Intelligence.

KC-10 – Tanker aircraft, equivalent to DC-10 in commercial fleet.

KC-135Q – Tanker aircraft modified with special ranging equipment for SR-71. Civilian equivalent is Boeing 707.

LPU – Life Preserver Unit.

MAC – The now defunct Military Airlift Command, comprised of Air Force transport aircraft. Responsible for moving cargo and people globally.

max/min – maximum/minimum.

mic – microphone.

Mobile – The vehicle which always helped launch and recover the SR-71. Normally manned by another SR-71 crew, who could communicate with the jet via installed radio.

MRS – Mission Recorder System. A tape onboard the SR-71 which recorded numerous bits of data during flight. Invaluable for reviewing by maintenance personnel.

OBC – Optical Bar Camera. Large nose-mounted camera in SR-71.

ops officer – Operations officer, normally the second in command in a squadron.

PSD – Physiological Support Division. The group responsible for maintaining the space suits. Also helped crews with suit-up and strapping into the cockpits.

recce – Reconnaissance. Useful term for pilots who have trouble with big words.

Redflag – Air Force war game exercise at Nellis AFB, Nevada. Flying units from around the Air Force and Navy participate throughout the year in simulated aerial combat. Some of the most challenging and intense training missions pilots will ever fly .

RSO – Reconnaissance Systems Officer. The navigator in the back seat of the SR-71, performing a host of duties besides navigation.

SAC – Strategic Air Command, now defunct. Comprised primarily of Air Force strategic bombers and intercontinental ballistic missiles. Was the home for the SR-71.

SAM – surface-to-air missile.

SAS – Stability Augmentation System. Intricate system which helps high performance aircraft maintain stability and optimum handling characteristics through damping of control movements via computerized signals.

sim – Simulator. The aircrew training device from hell (but very valuable for training).

Skunk Works – Nickname given to Lockheed's Advanced Development Projects division at Burbank, California. The secret plant was near Lockheed's plastics shop, and the strong odor permeated the plant, causing employees to joke that it resembled the infamous skunk works of the Li'l Abner cartoon fame.

Sled Driver – Title of the award-winning SR-71 book by Mach 1, Inc. Already owned by thousands of really cool folks interested in the SR-71.

sortie – A flight, regardless of duration, consisting of a takeoff and a landing.

spikes – The large cone-like devices in each inlet of the SR-71. As the aircraft accelerates past Mach 1.6, the spikes move aft in unison to position shock wave optimally for high speed flight. This is accomplished automatically through the DAFICS system, but the pilot has controls to run spikes manually if necessary.

TAC – Tactical Air Command, comprised primarily of Air Force fighter aircraft. Now called the Air Combat Command.

TACAN – Tactical Air Navigation. System of navigation used primarily in military jets, using radio signals from an emitting beacon to receive bearing and distance information. Less accurate than an INS, but fairly reliable.

TDI – Triple Display Indicator. Critical gauge, in both cockpits of SR-71, giving Mach, equivalent airspeed, and altitude readout. Especially necessary when flying at speeds and altitudes which rendered normal pressure-impact gauges unreliable.

TDY – Temporary Duty. All SR-71 crews were assigned to Beale AFB, in California, and going to DET 1 or DET 4 for their normal six week rotations constituted a TDY.

TEB – Tri-Ethyl Borane. Chemical substance used as catalyst to ignite the low-flammability JP-7 fuel during engine start and afterburner lights. It ignited on contact with oxygen.

tech rep – Technical Representative.

TEOCS – Technical Objective Cameras. Housed in central bays of SR-71 providing extremely sharp images. Normally used to back up other primary sensors. Could be run automatically through ANS signals.

The Ranch – Nickname for the Groom Lake area in the remote Nevada desert where the first Blackbirds were flown and tested. Also referred to as Area 51. Very secretive flight research center.

the 'take' – The pictures or other data brought home from a recce flight.

UCD – Urinary Collection Device. Don't ask.

unstart – Condition, at high Mach, where due to any number of adverse conditions, the shock wave is expelled from the inlet, causing spike to move forward to regain it. With one spike aft, and the other forward, a severe yaw condition is created which the pilot must correct immediately.

WSO – Weapons Systems Officer. The navigator in the back seat of the F-4 or the right seat of the F-111, performing navigation and weapons employment duties.